Landscape Fascinations and Provocations

READING THE AMERICAN LANDSCAPE

Lake Douglas, Series Editor

WITH CONTRIBUTIONS FROM

James Curtis
M. Elen Deming
Carol Emmerling
Rosa E. Ficek
Sharon Harkness
Terry Harkness
David L. Hays
Kenneth Helphand
Randolph Hester
Gary Hilderbrand
Lewis D. Hopkins
Misa Inoue
John Jakle
Doug Johnston
Rachel Leibowitz
Reuben M. Rainey
Robert B. Riley
Achva Benzinberg Stein
Frederick Steiner
Chip Sullivan
Linnaea Tillett
Suzanne Turner
Michael Van Valkenburgh
Vera Vicenzotti
Sue Weidemann
James Wescoat
Joachim Wolschke-Bulmahn

Landscape Fascinations and Provocations

READING ROBERT B. RILEY

Edited by Brenda J. Brown

Louisiana State University Press   Baton Rouge

Published with the assistance of the V. Ray Cardozier Fund

Published by Louisiana State University Press
lsupress.org

LSU Press Paperback Original
Manufactured in Canada
First printing

Designer: Michelle A. Neustrom
Typefaces: Whitman; Source Sans Variable
Printer and binder: Friesens Corporation

Cover illustration: Robert B. Riley, Illinois. Photograph by Paul Lettieri, 1975.

Essays by Robert B. Riley are reproduced with the permission of the Riley Estate and the original publishers, as follows:

"Editorial Commentary: Some Thoughts on Scholarship and Publication," *Landscape Journal* 9, no. 1 (1990): 47–50, © 1990 by the Board of Regents of the University of Wisconsin System, reprinted courtesy of the University of Wisconsin Press; "What History Should We Teach and Why?" *Landscape Journal* 14, no. 2 (1995): 220–25, © 1995 by the Board of Regents of the University of Wisconsin System, reprinted courtesy of the University of Wisconsin Press; "From Sacred Grove to Disney World: The Search for Garden Meaning," *Landscape Journal* 7, no. 2 (1988): 136–47, © 1988 by the Board of Regents of the University of Wisconsin System, reprinted courtesy of the University of Wisconsin Press; "Square to the Road, Hogs to the East," *Places Journal* 2, no. 4 (1985): 72–79; "Riley's Auto Territoriality," *Landscape*, Spring 1968, pp. 1–2, MSS 633 BC, J. B. Jackson Papers, Center for Southwest Research, University Libraries, University of New Mexico; "The Goose and the Dish" in *The Camaro in the Pasture: Speculations on the Cultural Landscape of America*, 154–57, © 2015 by the Rector and Visitors of the University of Virginia, reprinted by permission of the University of Virginia Press.

Library of Congress Cataloging-in-Publication Data
Names: Brown, Brenda (Brenda Joanne), editor.
Title: Landscape fascinations and provocations : reading Robert B. Riley / edited by Brenda J. Brown.
Description: Baton Rouge : Louisiana State University Press, [2023] | Series: Reading the American landscape | Includes bibliographical references and index.
Identifiers: LCCN 2022049662 | ISBN 9780807179321 (cloth)
Subjects: LCSH: Riley, Robert B.—Influence. | Landscape architecture. | Landscape design. | Cultural landscapes.
Classification: LCC SB472.3 .L36 2023 | DDC 712—dc23/eng/20221117
LC record available at https://lccn.loc.gov/2022049662

In memory of Bob

For those whose lives and thought he touched
and for those students of landscape and design who never encountered him

Eschew Obfuscation

—Bumper sticker, late 1980s

Contents

IV. Gardens, Technology, Fantasy, Experience

V. Landscapes and the Road

VI. Reflections on Pasts and Futures

VII. Reminiscences, Anecdotes, Assessments

Landscape Fascinations and Provocations

Robert B. Riley outside Mumford Hall, University of Illinois at Urbana-Champaign, 1990s. (Photographer unknown, Riley family collection)

times personal advice. Apparently, the drive and ability to understand other people and respond to their ideas that Illinois colleagues saw also shaped Bob's involvements beyond that university. They certainly served him well as editor of *Landscape Journal,* a role he assumed with little fanfare following the editorship of the journal's founders, Arnold Alanen and Darrel Morrison. Bob's previous involvements prepared him well for choosing reviewers for the wide-ranging submitted manuscripts. He usually knew whose specialty would fit and who would take time to review a particular manuscript well. However, while a substantial part of an editor's job is to process, improve, prepare, and publish others' articles, I also think—and Bob would agree—that *Landscape Journals* from this period show his stamp in additional ways. He announced his ambitions when I began working with him (as editorial assistant the first year, then as assistant editor). First, he wanted conference and exhibition reviews; among my tasks was to solicit writers for and write some of the first of these. I relished the opportunity and encouragement to analyze, contextualize, and reflect—not simply report—in writing them. Bob also wanted a "Most Important Books" feature with input from a wide range of people in landscape architecture and related fields. This piece, along with our subsequent "Most Influential Landscapes" and "Most Important Questions" (although viewed by some as elitist), brought intelligent, diverse voices to the journal and still make for interesting and enlightening reading. I cannot know how *Landscape Journal* was received "out in the trenches" then—my view was very much from the inside—but I believe it was noticed, by some with enthusiasm, by some perhaps with consternation. This was a time when print publications still ruled.

Bob's position as editor likely underscored his intellectual influence, especially on younger scholars. He was interested and willing to publish on controversial topics. He welcomed features such as "Landscape Architecture and Critical Inquiry" and devoted most of an issue to articles from the Women Land Design symposium—and he contributed to them himself. When developing his article "Must Landscapes Mean?" Marc Treib remarked that he wanted to publish it under Bob's *aegis.* Maybe Treib's usage was purely figurative, but I suspect for others that aegis maintained more of the mythological overtones of its etymology. Maggie McAvin, then a professor at Rhode Island School of Design, once told me she thought Bob did not realize how much he was a mentor to younger colleagues, herself included.

One can understand a lot about Bob by reading what he wrote. He read widely and he wrote well. His voice remains—lean, straightforward, erudite, wryly observant, provocative, and dense—dense in the sense of richness of ideas and depth and organization of thought. Over his career his writings included studies and recommendations for North America's cultural landscape—New Mexico rural villages, the northern plains, Midwest grain elevators, the strip—as well as contemplations on landscapes less specific, landscapes as variously experienced and portrayed—landscapes in literature, in memory; landscapes as images, as imagined; landscapes and gender; the garden. . . . These more philosophical pieces include organized surveys and discussions of existing relevant literature, as well as clarifying definitions, frameworks, criticisms, speculations, questions, and recommendations. Yet, while his writing was informed by and acknowledged diverse fields, Bob never failed to bring his arguments back to landscape, landscape scholarship and design, and even in his early work in *Landscape,* he displayed a propensity for calling designers to account.

Much of his writing is peppered with witty, occasionally scathing, *bon mots,* even aphorisms. "Rhetoric routs reflection once again."[3] "Sudden enthusiasm is no more valid a guide to merit than fashionable disdain."[4] "But if designers are to accept the charge of elite leadership they should look back once in a while to see if anyone is following."[5] "The scholarship that will help us is a far cry from the current deluge of picture books and from the model of garden history that delves ever more deeply into archival minutiae."[6] ". . . anticipation and memory are dialectic partners in our internal landscapes."[7]

There are many paradoxes and ironies to Bob and his career. Born and raised in Chicago, his landscape investigations focused primarily outside cities. With two baccalaureate degrees, he worked hard to launch Illinois's PhD program in architecture and landscape architecture and was honored and pleased to hood its first graduate. However, even as he worked toward it, he worried about such programs' potential effects on scholarship and academic subculture and later was dismayed to see some of his fears manifested. Bob was well aware of these ironies. Sometimes they made him uncomfortable; sometimes he enjoyed them. Referring to our work on *Landscape Journal* he once gleefully muttered, "Not bad, for an architect and a sculptor."

I suspect, with his abilities and talents, Bob would have left his mark in any time, but I also think Bob was very much *of* his time. He had what might be called a fortuitous entrance. He "fit" enough for acceptance, even leadership, but was different—curious, analytical, reflective, articulate, caring, and simply *smart* enough—to enrich, if not change, the field and people in it. His writings, with their many frameworks, insights, questions, analyses, and criticisms, continue to be relevant and often important. Bob helped others—not just his students—look at the world, or worlds, differently—more clearly, more richly, more consciously. He changed lives.

About the Essays

Robert Riley's publications span more than fifty years. Whether in speech or in written word, Bob strove to be clear and straightforward—especially about landscapes, their study and design. His essays in this volume suggest the breadth and depth as well as the ongoing relevance of his concerns. Some of those included also appeared—sometimes revised—in *The Camaro in the Pasture* (2015). However, for various reasons, I have chosen to reprint them here as originally published.

While Riley always brought his considerations back to landscapes and design, he conversed with anthropologists, geographers, and sociologists as well as designers and planners. He knew much of their work and found meaningful ways to test and integrate some of their thinking with his, always tempered with a strong dose of common sense. And so, while some of these new essays' authors are landscape architects, there is also a historic preservationist, a regional planner, an anthropologist, and an environmental psychologist/lighting designer. Their prior knowledge of Riley ranged from a comprehensive grasp of his work to vague name recognition. All these essays were written specifically for this publication. Each author was charged with the task of writing an essay that somehow related to Riley's work.

The essays are grouped loosely by theme, Riley's essays creating the scaffolding for those of the others. "Some Thoughts on Scholarship and Publication" (1990), Riley's first in this collection, was written fairly early in his tenure as editor of *Landscape Journal*. He probes the character and quality of landscape architecture discourse, articulating the difference between research and scholarship, arguing against "sloppy concern" and for "rigorous thought," himself providing an example of the latter in the process. While the concerns Vera Vicenzotti describes in "Landscape Architecture and the Scientific Method: IMRaD and Alternative Ways of Writing and Thinking" have obviously been brewing for some time, Vicenzotti was also inspired by Riley's example to examine the appropriateness, validity, and implications for current standardized research methods that often structure landscape inquiries and their presentations. In her "More Important Questions," M. Elen Deming refers to Riley's 1990 piece as an instance when "the editorial gloves came off completely," and "Riley threw down the editorial gauntlet in defense of rigor." As her title suggests, Deming's primary intention is to consider, reconsider, and reflect upon the "Most Important Questions" feature that ran in *Landscape Journal* in 1992 and to speculate on what those questions are today. However, she goes well beyond that; she is the author that grapples most explicitly with the time, milieu, and climate in which Riley's influence was most broadly felt.

Deming refers to the next Riley essay included here, "What History Should We Teach and Why?" (1995), as a polemic. As is typical of Riley, it is filled with ideas, telling comparisons and analogies, and pragmatic and provocative observations and suggestions. Riley lays out six "personal commandments for any history we teach," and three alternative directions that landscape architecture history courses might take, depending on how the landscape architect is viewed. Seeing the landscape architect as form giver would lead to a history of form, the landscape architect as a professional embedded in society would lead to a social history, and the landscape architect as a manager of change on the land would lead to a history of landscape change. In "Seeing the Unseen and Listening to the Local," Rachel Leibowitz, whose PhD committee was chaired by Riley, examines projects and processes in her work as a landscape preservationist in light of lessons and principles pertaining to history, landscapes, and change that she takes from this Riley essay (as well as from her many other interactions with Riley).

At his most reductive, Riley asserted that landscapes arise from a natural base, human culture *and* technology, and his sensitivity to technology's influence on landscapes' form and experience is apparent in the rest of his essays here. "Square to the Road, Hogs to the East" (1985) is perhaps one of his best-known writings on what he would call the ordinary American landscape, most specifically the landscape of rural east central Illinois. As the title—surely one of his most memorable—suggests, Riley's discussion is built on field observations and conversations, reports and statistics from the US Census Bureau and *The Prairie Farmer*, and, it seems, his own experience driving a combine. He is especially concerned with the technological forces that have affected this landscape "in complex and not always obvious ways." In the two essays that follow, Lewis D. Hopkins and I less sweepingly consider two different rural landscapes of today. Hopkins, in "Two Hundred Years of a Farm Landscape," responds to Riley's 1993 "The New Rural Landscape," which highlighted changes in the people and appearance of rural lands, particularly in east central Illinois. Countering Riley, Hopkins, with a nod to Kevin Lynch, celebrates the continuities amidst change in rural north central Ohio, his focus a farm in his family for over 180 years and its entwinement with the community around it. Although I refer to Riley's "Square to the Road, Hogs to the East," I think of my own "Snapshots and Fragments: Manitoba Farmstead Shelterbelts and Their Stories" as much an homage to the varied landscape portrayals Riley loved and used in his teaching as it is a response to his writing. Very much a visual *and* verbal documentation, it focuses on seven Manitoba farmsteads, their shelterbelts, and the people who live with them. It offers a preliminary glimpse into rural Manitoba life and landscapes today.

My first reading of Riley's "From Sacred Grove to Disney World: The Search for Garden Meaning" (1988) was something close to thrilling for me. Its subject appealed to me, but I was also gripped by the writing, the sort of bold wrestling to the ground of a complex subject. In classic Riley fashion the writing is clear, dense, sweeping, witty, dry (occasionally wry), observant, and demanding. "This essay is a simple and naïve pursuit of meaning in the garden" it begins, but it is simple and naïve only in the sense of probing fundamentals. Here Riley also discusses relationships between scholarship, research, and design, declaring that "each of our garden designs should pose a hypothesis and a plan to test it." His "few simple rules [to] help in our search for meaning" include the provisos that we acknowledge, not flee, technology, and that we build on gardens' capacity to provoke fantasy. In "Luminous Technologies and Nighttime Garden Enchantments" Linnaea Tillett and I discuss changing technologies of garden lighting, especially contemporary ones, and their potential role in garden experience and fantasy. Although nighttime garden experience was outside the purview of Riley's essay, I've little doubt he would welcome such an examination of this realm and its fantastical potential.

"Auto Territoriality," the earliest Riley essay included here, first appeared in 1968 in the Notes and Comments section of *Landscape*, where Riley was architecture and city planning editor. Riley applied the insights of Edward T. Hall to explain part of automobiles' appeal and people's resistance to public transportation. "Put simply," he writes, "the automobile allows one to travel almost anywhere in the public domain while remaining in a completely private world unequivocally defined by physical boundaries." As portrayed by Rosa E. Ficek in "The Overlanders: Technology and Culture on the Pan-American Highway," such autoterritoriality is integral to the lives of today's *overlanders* traveling the Pan-American Trail. Ficek's account of historical precedents of such privileged, insulated, long-distance travel, experiences of landscapes of and on the road, and of the Pan-American Highway itself, informs her discussion of this contemporary phenomenon and the electronic technologies and communications that facilitate and shape it.

Riley's "The Goose and the Dish" appeared in 2015, some twenty years after the most recent of his other essays included here. It is one of the shortest, most sweeping and reflective. Riley draws on his fifty-plus years of looking at, pondering, and teaching about landscapes. It is the penultimate essay in *The Camaro in the Pasture,* followed only by his "Closings." His dry wit is once again apparent. Acutely aware of the changes new technological developments have wrought to nonurban landscapes and their experience, he writes, "The new landscape is a *network,* based on different motivations, economics and sociology. It is a network with many fewer locational, more spatial distance restrictions than the old network and, in fact, with the electronic communications about as aspatial as any phenomena in space could be." The new landscape is "a very loose net laid down over, but almost independent of, the old landscape. Interaction between the two takes place at infrequent and unpredictable junctures, or nodes or linkages."

In his acknowledgments at the beginning of *The Camaro in the Pasture,* Riley cites several people with whom he "spent countless hours, and more than a few late nights, in animated, sometimes heated, encounters." Achva Benzinberg Stein is one of those cited. In the spirit of such encounters, here Stein casts her essay "The Environment Is a Public Good" as a continuing discussion with Riley. She takes issue with his vision of a future landscape in "The Goose and the Dish" by examining evolving ideals and realities of the public good, particularly as related to the environment and the role—and moniker—of landscape architects.

In one way or another, all the essays here demonstrate the continuing relevance and importance of Riley's larger view and way of looking at the world. His own writing stands strong on its own, and others' essays—each worthy of publication and attention in and of itself—reveal yet more facets. There are analytical probings, reflections, and explorations of changing everyday and high-end landscapes, their experience and design. There are philosophical arguments, calls to conscience and action, and strivings and discoveries of the fanciful and poetic. I like to think these essays would provoke Bob to exclaim "that's fascinating!" or "I never thought of that!," as he would when captivated by a new idea or perspective to ponder.

NOTES

1. Lewis Hopkins, personal communication, 2019.

2. Robert B. Riley, "Educational Accomplishments" (unpublished, 1992).

3. Robert B. Riley, "Gender, Landscape Culture: Sorting Out Some Questions" (*Landscape Journal* 13, no. 2, 1994): 160.

4. Robert B. Riley, "Reflections on the Landscapes of Memory" (*Landscape* 23, no. 2, 1979): 12.

5. Robert B. Riley, "From Sacred Grove to Disney World: The Search for Garden Meaning" (*Special Issue: Nature, Form and Meaning,* Anne Spirn guest editor, *Landscape Journal* 7, no. 2, 1988): 143.

6. Riley, "From Sacred Grove to Disney World," 142.

7. Robert. B. Riley, *The Camaro in the Pasture: Speculations on the Cultural Landscape of America* (Charlottesville: University of Virginia Press, 2015), 160.

Reuben Rainey, Achva Benzinberg Stein, Robert B. Riley. Annual Conference, Council of Educators in Landscape Architecture, University of Virginia, 1992. (Courtesy CELA and LAF)

I. Landscape-Based Practice, Scholarship, the Academy

Some Thoughts on Scholarship and Publication

ROBERT B. RILEY

This is my first issue as the sole editor of *Landscape Journal*. It would be conventional to offer platitudes about challenges, opportunities, and exciting times for the discipline. Instead, I will raise some thoughts that have been troubling me about the context in which this and other academic environmental design publications operate. I have within the last year participated in four conference sessions discussing the current state of scholarship and graduate education in environmental design. The major concern of the educators at these meetings can be expressed simply: raising the standards of scholarship and research, while maintaining both a diversity of approaches and some sense of relevance to real-world design—the bettering of the human environment. What follow are fragments and pieces of my own concern.

On Scholarship and Research

Research is a magic word. Its effects on the discipline have been mixed. On the one hand, there is a pride in our doing research, a pride that we are part of the scientific and intellectual community. On the other hand, there is a sloppiness about the use of the word that produces confusion about what we are doing and why. The term *research* is used too broadly, surely. At one extreme, most of us recognize a narrow model of research from the experimental or laboratory sciences: the posing of a hypothesis, the designing of a method to test the hypothesis, the testing itself, conclusions, modification of the hypothesis, and so on. At the other extreme, our undergraduate design students, flipping through magazines the first week of a design project, are "doing their library research." Somewhere in the magic of research, the idea of scholarship has gotten lost.

Is this just a semantic quibble? Beyond the vague feeling that *research* and *scholarship* are not identical, is anything to be gained by examining how the two words are used? The recent social etymology of the words is certainly distinct. Research has not only connotations of laboratory and experimental sciences, it has a mystique that has pervaded American society and universities since the Manhattan Project. Research is glamorous. Research produces bigger bombs, faster Fords, and cancer cures. Scholarship produces intellectual masturbation on the use of verb forms in Chaucer. Real folk do research, wearing hard hats and laboratory smocks. Quiche folk do scholarship, wearing cardigans and tennis shoes.

Research has even become a verb (though not in the pages of *Landscape Journal,* thank you). Can we imagine a verb *to scholar?* Imagine the ASLA Annual Awards Program including scholarship awards as well as research awards? Imagine referring to our institutions as major scholarship universities? Not likely. Does it matter? Yes, I think it does. First, precision in the use of words is never amiss. Second, the rush to use the word *research* to cover almost anything that isn't teaching or service leads to sloppy thinking and sloppy standards.

Finally, I think the implicit tendency to model ourselves on quantitative, or laboratory, or experimental, or whatever, sciences has even affected the format of what we write, and thereby how we think. Two small examples are the use of citations and abstracts. In the experimental sciences, citations are commonly used to base work upon earlier experiments in the same line. In these fields, one commonly assumes that the findings referred to have been "proven." In the speculative, humanist tradition, this is very different. A reference to Jay Appleton, for example, does not prove that human beings react to the environment in terms of prospect and refuge; it only establishes that Appleton suggests they do. And yet I constantly get manuscripts from authors who don't know the difference. This is the phenomenon known as "authority by repetitive citation." It is a good example of low standards of thought in our discipline.

Similarly, the role of abstracts is different in laboratory sciences and in a field such as ours. In most established hard sciences, an abstract can give a fair idea of both the meaning and the quality of a paper, unless the author is either negligent or fabricating. This is because the fields are narrowly focused and paradigms are well defined and universally accepted. In a field such as ours, an abstract conveys no idea of the quality of the paper and often little even of the content. One of the marks of our maturity as a discipline will be when CELA, for its annual conference, insists on reviewing complete manuscripts and not abstracts.

On Professions and Disciplines

Academic landscape architecture is in transition from a profession to a discipline. The concerns and the activities of our academic world are no longer driven by the world of practice and applications. For years, academics in the applied and fine arts have been accepted as members of the academic community without being expected to conform to its standards. Environmental designers within the academy did essentially the same things, or addressed the same issues, as those in the outside world. Practitioners faulted the schools for how well or poorly they prepared students for practice, but there was never any question that that was what the universities were doing, however well or badly. That time is passing, maybe passed. Academic landscape architects are doing things that are not done in practice. Some of these, like environment-behavior studies, are unique to academia because the world of practice has not become savvy enough to realize it should be using them, or because the state of knowledge has not advanced to the point that such studies are usable in actual design. Other activities going on in the universities, like the scholarly study of landscape architecture itself, will probably never be done in practice. This is only an aspect of the change, however. Landscape architectural educators are becoming less like their practitioner cousins, more like their academic colleagues. The academic design world is becoming a subculture of its own, distinct from the world of practice. The growth of this separation, which the classic modem movement fought to avoid, is a subject for study in itself as an exercise in the sociology of disciplines. But it is certainly clear that the academic design community now communicates largely with itself, has established its own set of awards and peer recognition, socializes its children, provides rites of passage, and above all poses its own questions and develops its own paradigms distinct from, and sometimes irrelevant to, the world of practice. This is as much a cause of the lack of communication between the practitioners and academics as it is a result. It is not conformity forced on us by academic administrators. Our constraints come less from administrators than from ourselves and our peers and are reinforced by promotion and tenure hysteria. The questions we seek to answer are not only distinct from those in practice, but they will undergo cycles of fashion all their own, equally remote from, and irrelevant to, practice. For indeed there is a cycle of fashion in academic concerns as much as there is in design styles. Whether we view all this as maturation or degeneration, as problem or opportunity, it seems inevitable.

Other fields have been there before us. Different fields have developed different relationships between study and application, between discipline and profession. The form of the distinction in medicine is not the same as in law, and neither is the same as that in music. Such speculation can be amusing but also frightening. Could environmental design be headed to an eventual split as radical as that between English and journalism in most universities?

On Magazines and Journals

If these traits distinguish between two sociological subcultures, the issues are not so clear intellectually or ideologically. There are a number of people out there, subculture or not, who think they are doing something akin to design "scholarship." I suppose they think of themselves as design theoreticians. They are a mix, ranging from art and architectural historians to academic teachers of design theory to high-style practitioners who appear on the cover of *PA* or *LAM* and issue theoretical pronouncements. Often these folks have a loose, half-time, adjunct relationship to the academy. Much of the visual, spatial, and even intellectual excitement in the design world comes from them. But yet can we classify them as academics or practitioners only? Think about publications. We all know that practitioners have magazines and scholars have journals. On this basis, it's easy to distinguish between, say, *Progressive Architecture* on the one hand and the *Journal of Architecture Planning Research* on the other, between *Landscape Architecture* magazine and *Landscape Journal*. But there are influential periodicals that do not easily fit into this dichotomy of professional, ad-filled magazines versus scholarly journals. How about, for example, *Space and Society, Oppositions, Places, Landscape?* They commonly publish provocative and speculative musings of an orientation and quality that seldom show up in the conventional magazines or conventional journals. It is an exercise in these distinctions, or lack of distinctions, to think how you might decide between sending an article to this journal or to *Landscape* or *Places*. This is important not only because we live in a media dominated society where so much of our intellectual communication takes place in print, but also because much of our best philosophy, criticism, and speculation come from these "in between" publications.

On Theory and Pseudotheory

Theory probably shares with *sense of place* the award for the most misused term of the decade in environmental design. There are two important differences between the use of these two terms, however. First, *sense of place* will no doubt pass out of fashion before too long. The word *theory* has been around a long time and probably always will be around. The other difference is that sense of place is a real-world phenomenon, of acknowledged importance if of difficult definition, and is, of necessity, a fuzzy set. Theory, however, is a concept and a definition and can be as discrete and unequivocal as we wish it to be. Unfortunately, the definition of theory as landscape architects use it is simple. *Anything concerned with what to do or why to do it, instead of how to do it, is proudly proclaimed as theory. This is not theory; this is pseudotheory.* In addition to this broad and sloppy definition, pseudotheory is characterized by two other properties.

Pseudotheory Is Plagiarized from Other Disciplines. Such justificatory borrowing goes back at least as far as Palladio, but has been a dominant theme of architecture from Viollet-leDuc through Modernism to Post-Modernism. John Summerson put it well four decades ago:

> The Modern Architect . . . has, for some reason or another, stepped out of his *role*, taken a look at the scene around him and then become obsessed with the importance not of architecture, but of the *relation* of architecture to other things. This is exactly what has happened. The architect has walked out of himself, rather like a second personality is seen to walk out of the first in a psychological film. He has (to pursue this metaphor for a moment) left the first personality at the drawing board and taken the second (the "live" personality) on a world-tour of contemporary life—scientific research, sociology, psychology, engineering; the arts and a great many other things. Returning to the drawing board he finds the first personality embarrassing and profoundly unattractive. There he stubbornly sits, smelling slightly of "the styles." So the second personality sits down beside him and painfully guides his hand. (P. 197)

Thus we have had banal biomorphism (a.k.a. "organic"), pseudosemiology, and now decorative deconstructionism. We also currently have flatulophenomenology; but to give it justice, phenomenology as a philosophy or an epistemology, not a theory, and as used by Europeans, not North Americans, has made serious contributions. Readers interested in thinking about this topic should read Summerson (1963) or Peter Collins (1965). I will mention only briefly my discomfort with one currently most fashionable mode of pseudotheory. Deconstructionism is a theory of literary analysis. The simplistic inference made by our high-style design pundits—that a complex of theory and methods designed to investigate a literary text is equally valid for the built environment—is an example of the superficiality and lack of disciplined thought that pervades the design world. In fact, serious discussion about ways in which buildings and landscapes are similar to literature in the perceptions of their consumers and ways in which they aren't would be thought provoking. The mind-jerk assumption of their equivalence is but nonsense. It typifies our inability, or unwillingness, to distinguish model from metaphor, analysis from analogy, theory from framework.

Pseudotheory Follows Form and Seeks to Justify It. Styles can be thought of as having their own life cycles, susceptible to study apart from the society in which they are embedded. Many historians of art have taken this approach. Proponents of a new design style, however, particularly in situations where its acceptance depends on publicity, justify it in terms of larger social or intellectual profundities. An extreme example of this was heard at a recent Associated Collegiate Schools of Architecture meeting. The familiar Post-Modern cliché of intersecting grid systems producing perceptual ambiguity (Tschumi, what have you wrought?) was justified on the basis of what we have "learned" from science, more specifically, physics. Einstein's theory of relativity and Heisenberg's uncertainty principle were used both to justify the particular design solution and to discredit classic Modernism as a valid aesthetic vocabulary for our time, Einstein and Heisenberg having proven the futility of a unitary interpretation. The irony, unmentioned at the conference, is that the bible of classic Modernism, Gideon's *Space, Time, and Architecture,* used exactly the same argument to justify the Modernist architectural revolution and discredit the Beaux Arts eclecticism which preceded it. We should be able to laugh at this, but so many design schools take such silliness so seriously that the laughter rings hollow.

On Theories, Models, and Frameworks

A plea for more restrictive and more precise use of the word *theory* is not in conflict with the plea for diversity of methods and approaches. Precision, per se, is never out of place. Further, by making clear what we are doing and not doing, restricted use of the term *theory* should, in fact, support a diversity of approaches, a point to which I shall return. The definitions that I find most useful in this arena are those of Amos Rapoport, who distinguishes between frameworks, models, and theories. Put in simplest terms, a theory explains, a model predicts, and a framework organizes. A framework can be judged on its reasonableness and its utility, but claims no exclusivity vis-à-vis other frameworks. If we accept these definitions, it is clear that most theories or philosophies of design are not theories at all, but frameworks. Philosophies of design are intellectual and verbal frameworks that organize ideas, just as styles can be thought of as visual frameworks that organize tangible things. Design theory or philosophy, then, is often a sociocultural framework serving as a justification for a visual framework. Aphorisms form a fourth, and I think useful, intellectual category. It is likely that the design world has used aphorisms more than other fields, but that is another topic. What is clear is that expressions such as "form follows function," or "architecture is frozen music," or "less is more" are shorthand selling slogans for visual frameworks or styles.

This takes us back to the second advantage of using a restrictive definition of theory. By distinguishing among the four classes of aphorisms, frameworks, models, and theories, we clearly see what stage of intellectual certainty or utility we have established with any given statement. But there is also a curious fringe benefit. Some famous aphorisms, maxims of design, are in fact translatable to theories capable of being tested. Of the three examples quoted above, one—"less is more" falls most obviously into this category. We can paraphrase Mies's maxim as "economy in formal expression produces greater affective impact upon the beholder." (That there are other equally reasonable translations points to a critical difference between an aphorism and a theory, and partly explains the persuasive power of the former.) A social scientist could test this generality in specific, controlled experiments. Maybe the relationship between aphorisms and theories can be seen as another expression of the relationship between the "aesthetic" and the "behavioral" approaches to design—the critical design schism of our time.

Surely it is clear that the first step in building a valid, useful theory of design is understanding what a theory is. The second step is to realize that, while aphorisms might spring fresh born from someone's psyche, theories, at least nontrivial theories, do not. Theories are built, destroyed, and revised piecemeal. Theories of any utility or validity come from an interaction between inductive and deductive steps. Sandra Howell (1989) offers an analysis of this state of contemporary theory building in these terms:

> One of the unfortunate by-products of this anti-theory attitude is, ironically, production of instant "theory building." Believing that no one is looking back, academics in the environmental design–behavior research arena tend to presume that by naming something a "theory" they have cut a new swath in knowledge. Since few of us bother to publicly criticize such offerings (are we afraid of the awesome note of scholarship embedded in the word "theory"?), the culprits go unconfronted! Scholarship in Environmental Design must come to be an exchange between inductive (case generated) and deductive (theory generated) process. This is the *pic-up-stix* layering required for the culmination of knowledge that, hopefully, will, in fact, be useful to practice.

On Quantification of the Trivial for Confirmation of the Obvious

I know that many readers feel that *Landscape Journal* is dominated by a distinctive segment of our discipline, the fuzzy set, including behavioral investigations, visual assessment, land planning and analysis, and so on, which can be described as quantitative-analytical work. I can almost hear colleagues griping about the *Landscape Research* crowd, or the _____ clones (provide your favorite well-known author). I don't know whether this is true or not. Maybe it is something to be treated in a future editorial. But the belief bears on an issue worth discussing.

First, whatever editorial policy *Landscape Journal* has or has not, it cannot be explained by the conspiracy theory. While all the editors of this magazine have had their own personal inclinations and prejudices, I can say with some certainty that they have had little effect on the magazine content. If indeed, the segment described above has published out of proportion to its role in the discipline, the likely explanation is that the editors have received more of such manuscripts. People working within that mode of research have developed the habit of written publication as part of their education and their work. Other sectors of the discipline and the professions have not. But is it possible that the acceptance rate of such papers is higher? Again, I can't say, but I wouldn't be surprised.

People who submit quantitative-analytical manuscripts to *Landscape Journal* have usually received doctoral training in their specialized field or, at the very least, have trained extensively under people who have. People who submit quantitative work to this journal have learned to do research. Unfortunately, many of the people who submit "think pieces" have never learned to think. Anyone who takes offense at this harsh characterization should be forced to read some of the drivel submitted under the rubric of design philosophy or theory. There is a corollary to this situation: just as the authors of quantitative work have been trained within an accepted, well-defined paradigm, so have their reviewers. A quantitative research manuscript is easier, or at least cleaner, to evaluate for publication.

The hard fact is that the "humanist" or speculative tradition in landscape architectural scholarship will hold its own with quantitative research in this journal when, and only when, academic landscape architecture programs start teaching their students and their faculty the difference between sloppy concern and rigorous thought.

And . . . Finally . . . on *Landscape Journal*

Landscape Journal is a scholarly publication. If landscape architecture, and environmental design in general, is the provision of settings that functionally support and emotionally enhance the lives of people, then landscape architecture scholarship is the activity that provides the knowledge base for such design. The concept of knowledge base includes not only information, but intellectual structures that organize that information: frameworks, models, and, hopefully and eventually, theory. Being a scholarly journal implies the publication of the best in both quantitative research and the so-called humanist tradition. Specifically, this means that this journal should be running more design philosophy, criticism, and, yes, if you will, theory—done with the rigor and clarity of the quantitative work it runs alongside. If indeed there are two separate intellectual currents within our discipline, this journal should be representing the best of each. We could hope that they would begin to inform one another, eventually even affect one another.

REFERENCES

Collins, Peter. 1965. *Changing Ideals in Modern Architecture, 1750–1950*. London: Faber and Faber. See particularly chapters 14–17.

Howell, Sandra C. 1989, April. "Environmental Design: The Pic-Up-Stix of Academia." Unpublished paper circulated at the meeting of the Environmental Design Research Association, Black Mountain, North Carolina.

Summerson, John. 1963. "The Mischievous Analogy." In *Heavenly Mansions and Other Essays in Architecture*. New York: W. W. Norton.

Landscape Architecture and the Scientific Method

IMRaD and Alternative Ways of Writing and Thinking

VERA VICENZOTTI

Imperialistic Tendencies of Research

Due to the "fascination with the magic aura of research, the idea of scholarship has gotten lost."[1] This is one of the central observations that Robert B. Riley makes for academic landscape architecture in an editorial for *Landscape Journal*. With the term "research," Riley designates the mode of inquiry of the quantitative or experimental sciences. With "scholarship," he seems to designate the mode of inquiry in the arts and humanities. Describing both terms' "social etymology," Riley writes: "Research is glamorous. Research produces bigger bombs, faster Fords, and cancer cures. Scholarship produces intellectual masturbation on the use of verb forms in Chaucer. Real folk do research, wearing hard hats or laboratory smocks. Quiche folk do scholarship, wearing cardigans and corduroys."[2] The remainder of his editorial makes clear, though, that he doesn't share this view. In fact, one could read his editorial as *performing* rigorous scholarship, demonstrating its fruitfulness. Today, thirty years after the original publication of Riley's editorial, one could argue that there are signs that scholarship in landscape architecture has matured and gained reputation—I'm thinking of the rising number of PhD students in landscape architecture who are publishing scholarly (as opposed to research) articles (e.g., Danneels 2019; Jacobs 2019; Klosterwill 2019).[3] However, as Riley states in the notes of a slightly modified and updated version of the editorial in 2015: "the issues persist."[4] Research continues even today to exhibit imperialistic tendencies. This makes it difficult to perform, teach, and learn rigorous scholarship in landscape architecture.

With two brief examples, the use of citations and the use of abstracts, Riley illustrates how "the implicit tendency to model ourselves on quantitative, or laboratory, or experimental, or whatever sciences has affected the format of what we write, and thereby how we think."[5] In this essay, I would like to discuss one more example of how the lure of research undermines the idea of scholarship: the relation between mode of inquiry and text structure, that is, which outline or organizational scheme is chosen in a journal article, or thesis, or any other kind of academic text. A key example is the use of the IMRaD format as a seemingly neutral format to structure academic texts in landscape architecture. The acronym IMRaD stands for Introduction, Methods, Results and Discussion and describes the default structure of scientific texts. Believed to be the epitome of academic text structure, this format has been getting a preferential treatment in academic landscape architecture. Instances where this has happened may not be overly frequent; they are, however, symptomatic. They range from conferences where the IMRaD structure has been the requested format for papers, as anecdotal evidence has it, to teaching situations, in which the IMRaD format has been the standard structure recommended for bachelor and master theses, as I know from my own experience. My colleagues would argue that this structure helps students understand the nature of academic writing, that students appreciate the recipe-like character of the format, and that the structure is flexible and accommodating enough for any topic to be tackled. I disagree with all three points. In explaining my reasons for disagreement, this essay will shed some light on the relation between form, that is, text structure, and content of writing and ways of thinking in landscape architecture. In doing this, it will uncover one instance of "scientific imperialism"[6] in action in academic landscape architecture as well as present some alternatives to the IMRaD structure.

I will argue that the IMRaD model is not a neutral way to structure a text that suits all modes of inquiry equally. Rather, this format has historically evolved in the sciences and caters to the needs of this one particular mode of inquiry. Other disciplines have other traditions, conventions, and demands on text structure. Landscape architecture is an interdisciplinary field. So, elevating the IMRaD model to the only accepted text outline makes life difficult for or even undermines research, or scholarship, in the arts and humanities tradition.

My explorations are based on literature on the IMRaD structure from fields such as English or the history and philosophy of science, guides and handbooks on academic writing, as well as a qualitative analysis of the text structure of a few sample articles from *Landscape Journal* and the *Journal of Landscape Architecture* (JoLA). I consider these journals to be core journals in the discipline of landscape architecture. The themes, methods, modes of inquiry, and text structures of articles that appear in these two journals can thus be considered as belonging to (and forming) the ways of writing and thinking in landscape architecture. The selection of the analyzed articles is subjective and strategic: I focus on what I consider well-written papers within the arts and humanities tradition because I want to present and discuss established and viable alternatives to the IMRaD structure.

The IMRaD Structure

The IMRaD format is nowadays the standard outline for original research papers in many sciences, in particular experimental sciences and laboratory research, such as physics, chemistry, or biomedicine, as well as in landscape ecology.[7] It is also used in the social and behavioral sciences, when and if they work empirically, experimentally, and quantitatively. However, IMRaD has not been with us from time immemorial. It is the result of an evolutionary process and has a specific history.[8] Modern scientific writing has its beginnings in the seventeenth century, with authors such as Francis Bacon, Robert Boyle, and Isaac Newton. Authors of this period would write erudite letters to communicate their results. Over the course of the next century, this type of experimental report evolved to a more structured form. By 1800, many articles were organized by first outlining the overall "theory," then presenting a "specific hypothesis," and finally reporting about the "experimental trial-as-proof"; occasionally, conclusions could be found at the report's end.[9] During the twentieth century, this organizational scheme evolved further into the current IMRaD format. As Sollaci and Perreira show for journal articles in medicine, the IMRaD structure began to be used in the 1940s, reached 80 percent in the 1970s, and by the 1980s, became the only pattern adopted in original scientific research papers. In other fields, notably physics, this structure was extensively adopted as early as the 1950s.[10]

But why has the IMRaD format been adopted in so many scientific papers? According to Meadows, the emergence of highly formalized articles is a response to the constantly growing amount of information. Since the IMRaD organization determines what information goes into what paper section, it makes it easy for the reader to quickly scan the text and extract the searched-for facts.[11] Furthermore, as scientists themselves argue, it helps the author in writing a well-organized paper, and it facilitates the evaluation of manuscripts submitted for publication by reviewers and editors of academic journals.[12] However, there are other reasons for the enduring success of this organizational scheme, which are not likely to be written about in the myriad of instrumental handbooks and guides that instruct students and novice researchers on how to write scientific papers.[13]

Rhetorical and linguistic research on English scientific writing has shown that the IMRaD structure emerged as a solution to the rhetorical challenge "to gain agreement" from the reader:[14] The format helps to persuade an audience without direct access to the empirical events being reported that the findings described are accurate. Credibility is achieved by hiding the messy research process in the reported product: The "obviously idiosyncratic set of events" that constitutes experimental research is given "an objective appearance" "through the conventionalized 'packaging' of the research."[15] By highlighting the objective logic underlying the research presented, the IMRaD structure elicits credibility by strengthening the audience's impression that what is reported is (the result of) good science.

And what constitutes science? One easy answer is that it is "the scientific method," and the IMRaD structure mirrors this method. The International Committee of Medical Journal Editors, for example, describes the IMRaD structure as "not an arbitrary publication format but a reflection of the process of scientific discovery."[16] In textbooks and on educational web sites, the scientific method is often described as a well-defined procedure with some four or five steps, "starting from observations and description of a phenomenon and progressing over formulation of a hypothesis which explains the phenomenon, designing and conducting experiments to test the hypothesis, analyzing the results, and ending with drawing a conclusion."[17] We can see how the steps of "the" scientific method correspond to the sections of the IMRaD paper: the introduction presents the descriptions of the observed phenomena that prompted the research question and the formulation of a hypothesis; the material and method part explains the research design and methods used; the results of the experiment (or survey, field work, or any other data-generating equivalent) are described in the results section, while in the discussion part, the author draws conclusions that are formulated by interpreting the meaning of the results vis-à-vis existing research or a practical problem.[18] In this sense, the IMRaD method can be said to mirror the scientific method. Or in the words of Meadows: "In fact, the construction of an acceptable research paper reflects the agreed view of the scientific community on what constitutes science."[19]

However, the idea of *the* scientific method is quite contested in the philosophy of science; furthermore, scientific publications do not generally reflect the research process that led to the results.[20] Peter Medawar, Nobel Laureate in Physiology or Medicine, argued already in 1963 that the "scientific paper in its orthodox form," and by that he referred to papers adhering to the IMRaD structure, "does embody a totally mistaken conception, even a travesty, of the nature of scientific thought."[21] His argument was that papers with this format misrepresent the haphazard thought processes that characterize scientific work. This is true even today, when most scientific papers are presented as being hypothesis-driven rather than following an inductive logic.[22] Indeed, as philosophers, historians, and sociologists of science have argued, scientific publications have to be understood as *ex-post* reconstructions of the messy research activities.[23] There seems to be no easy answer to the question whether this is a good or a bad thing.[24] So, funnily enough, the advice given to students writing a philosophy paper seems to be good pragmatic advice for science students, too: "A good narrative order strikes a balance between a psychologically plausible order of discovery and an intellectually satisfying logical-explanatory order."[25]

Thus far, we have seen that the IMRaD structure has emerged within the sciences and caters primarily to the needs of experimental disciplines. But we have also seen that it is not uncontested—not even within the (experimental) sciences. And we will see in the next section that there are alternatives to this format, both within the sciences and within the context of other modes of inquiry.

Alternatives to the IMRaD Structure

There are at least two well-established alternatives to the IMRaD format. The first alternative, occurring within the sciences, is what one could term the "logical argument" structure,[26] or the "deductive plan."[27] Similar to IMRaD papers, articles that follow this format will start by citing observations, too, but then proceed by use of logic, established procedures, and suggestions for new procedures to decide what theory is relevant to explain the phenomenon in question.[28] This format seems to be common in astrophysics as well as other fields that do not lend themselves to experimentation and that are rather based on logical argumentation, such as theoretical linguistics or mathematics.[29] Furthermore, the IMRaD structure is rarely, if ever, followed in fields such as geoscience and engineering, where fieldwork and descriptive efforts tend to be central.[30] And last but not least, there are examples of epochal science papers that have defied almost any rule one will find in guidebooks on scientific writing, including adhering to the IMRaD outline. One famous example is Watson and Crick's "Molecular Structure of Nucleic Acids"[31] (see, e.g., the analysis in Montgomery).[32] One could rightly object that Watson and Crick's paper was published before the IMRaD format had risen to the standard outline. This objection is not wrong but misses the point, which was to show that the IMRaD format, while dominant in many scientific fields, is far from universal—even within the sciences. Beyond the sciences, different needs and demands have created different ways to organize academic texts.

THE TRADITIONAL ESSAY STRUCTURE

The second alternative, which is widespread in the arts and humanities, is the traditional essay. The typical essay consists of three major components: the Introduction, the Body, and the Conclusion. In the Introduction, the reader is introduced to the topic that will be discussed and, crucially, the thesis that will be argued for. The Body is the essay's main part. It is here that the "discussion is carried out and the results are presented."[33] We will see shortly that this inversion (by IMRaD standards) of discussion and results is no accident. In shorter pieces, the Body will consist of one section only but in longer texts, such as bachelor, master, senior, or whatever theses, it will itself be divided into several sections or chapters. In the Conclusion, the main argument will be summed up and closing thoughts will be presented. A traditional essay's structure may follow the model provided below.

I. Introduction
  1. background, context for topic
  2. transition to thesis
  3. thesis statement
II. Supporting point 1
  1. supporting detail
    i. example 1

ii. example 2
2. supporting detail
i. example 1
ii. example 2
3. supporting detail
i. example 1
ii. example 2
III. Supporting point 2
IV. Supporting point 3
V. Supporting point 4
VI. Conclusion
1. review central ideas presented in body and make connection to thesis
2. transition to closing thoughts
3. closing thoughts

Outline of the traditional essay (based on Excelsior Online Writing Lab, 2020)

However, the question remains: How can we structure the essay's body? What is the logic according to which we can order the supporting points, as they are referred to in the model outline? Pertinent guides will argue that there is no general answer to this question. Instead, one has to find, or rather: create the outline that, in the best possible way, will support the argument one wishes to make.[34] This means that each essay's structure is unique to the claim it makes. However, while there is "no set formula,"[35] certain types of solutions can be discerned.

Variants of the Essay Structure

For this essay, I could identify six different types of organizational schemes: the chronological outline, the comparative-contrastive outline, the analytical outline, the theoretically informed-close-reading outline, the spatial outline, and the cause-and-effect outline. Within the confines of this essay, I will examine the first four of them in more detail. I arrived at this list by starting from the overview Umberto Eco provides in his fine little book *How to Write a Thesis,*[36] combining it with suggestions in other guides[37] and my own observations, which are partly based on the analysis of sample articles and partly on my own experience of reading, writing, and editing scholarly texts in landscape architecture and neighboring fields. Thus, all of the outlines I will present shortly are fruitful and well-established in academic landscape architecture.

One of the most intuitive organizational schemes is the *chronological* outline.[38] Eco recommends it as a model outline for history theses, his example being "The Persecutions of Waldensians in Italy." A more landscape-related title would be "A Short History of Street Trees in Southern France." In the chronological outline, earlier events are presented before later ones. To the degree that history is more than chronology, a sequence of events, the different sections of a paper or chapters of a historical thesis can also assume the character of "layers of a cake, or . . . chronological sections along a line of argument."[39] They do not only recount what happened when, but also why and how—and what this means. A historical paper may, if it lays more emphasis on the latter questions, be considered a historically informed close reading of a site, a design, an oeuvre, or a landscape-related practice.

This is the case in Kenny Cupers's article "Gardening as Geopolitics."[40] This rather brief essay has four sections that, instead of having headings, are numbered with roman numerals—a practice that is not unusual in the arts and humanities. The first, fairly brief section, functions as an introduction. Here, Cupers prepares and presents his thesis statement, that twentieth-century history would have shown that gardening is not only a tactile individual experience but also that it has to be understood as a geopolitical act. In the following three sections, Cupers skillfully uses a chronological structure to back up and nuance his thesis statement. Kaleidoscope-like, each chronological layer shows the reader a new facet of gardening as a practice that is inextricably connected to imperial expansion ambitions, mass housing policies, and the transformation of domestic territory, revealing the complexity and geopolitical dimension of what is so easily believed to be an innocent and earthbound, purely individual practice.

If one is less interested in emphasizing causal relationships and more in thinking through commonalities and differences beyond first appearance, one might turn to the *comparative-contrastive* outline ("Nationalism and Populism in Italian Literature in the Great War Period," to cite again Eco's example). Classic compare and contrast texts, that is, comparisons that treat A and B equally, could either first discuss all of A, and then all of B (the block type), or they could alternate points about A with points about B (the point-by-point type).[41]

[Introduction]
Three Tyrannies of Contemporary Theory
Positivism
Paradigms
The Avant-Garde
Tradition and Hermeneutics
Hermeneutics
The Hermeneutic Landscape
Conclusion

Outline of James Corner's "A Discourse on Theory II: Three Tyrannies of Contemporary Theory and the Alternative of Hermeneutics" (Corner 1991)

James Corner's essay "A Discourse on Theory II: Three Tyrannies of Contemporary Theory and the Alternative of Hermeneutics"[42] is an example of the block type of the comparative-contrastive outline. In the first part of the essay's body, Corner presents and discusses three approaches to (landscape) architecture theory and what their tyrannies, or at least their downsides, are. This block is then contrasted in the body's second part by his presentation of the alternative: the hermeneutic approach. Between what one could define as the essay's body and its conclusion, Corner has placed a section titled "The Hermeneutic Landscape." In it, he sketches theoretically and through examples of actual landscape design how the hermeneutic approach to theory could work and look. This section functions as a transition to the conclusion, with which it could have been merged. However, by elevating it to the same outline level as the two other, much longer parts of the body, Corner emphasizes the importance of this section. It contains the proof, as it were, that the hermeneutic approach to landscape architecture theory is able to give truthful accounts of, as well as help, produce good landscape design.

Common, if not standard, in (analytical) philosophy, is a text structure that for want of a better term, I refer to as an *analytical* outline. Like pieces of a puzzle, the body's supporting details present the steps of the argument the author is making to support her thesis. The pieces are assembled to convince the reader of premises, or they anticipate and answer to objections to the advanced argument.[43]

[Introduction]
The Three Faces of the Picturesque
    Picturesque Style
    Picturesque Ideology
    Picturesque Aesthetics
The Role of the Imaginative Spectator
Appreciating the Ugly, Melancholy, and Grotesque
Distance from the Subject Matter
Conclusion

Outline of Susan Herrington's "Framed Again: The Picturesque Aesthetics of Contemporary Landscapes" (Herrington 2006)

One example of this is Susan Herrington's paper "Framed Again: The Picturesque Aesthetic of Contemporary Landscapes."[44] In this essay, Herrington holds that while recent works in contemporary landscape design may have left behind the formal and ideological aspects of the Picturesque, they have continued to participate in its aesthetic dimension—a thesis she introduces in the first, untitled introductory section. The text's body is divided into two major parts. In the first part, "The Three Faces of Picturesque," Herrington distinguishes between the Picturesque as style, ideology, and aesthetics. This distinction is both a contribution to knowledge and analytical tool. How the Picturesque as an aesthetic mode serves as guiding principle of current landscape design is "evidenced" by three points:[45] the primary role of the imaginative spectator; the appreciation of the ugly, melancholy, and grotesque; and lastly the distance from or unfamiliarity with the work's subject matter. Herrington explores each of them in turn in the body's second part. Herrington's essay, as well as all the other papers discussed, proceeds by blending what is neatly separated in the IMRaD format: results and discussion. In Herrington's piece, the three points of evidence are the closest equivalent to results in the IMRaD logic. However, they are not the result of passive and disengaged observation but of active and involved intellectual work—of interpretation. This is why it is detrimental, if not altogether impossible, for interpretative texts to present results without any discussion. Trying to rewrite, for example, Herrington's piece in the IMRaD format could be an instructive pedagogical exercise to prove this point.

An outline that tends to make the author's standpoint, their situatedness, very explicit is the *theoretically-informed-close-reading* outline (modeled after a type of text common in disciplines such as English literature).[46] Here, the object of the close reading—be it a site, design, oeuvre, discourse, or practice—is understood and read through the lens of a particular theory, for example, a postcolonial or a feminist theory, which is often announced very clearly. In texts with this type of outline, the reader will be introduced to the object of interpretation and then to the theory or vice versa; after that the actual interpretation follows: the close reading of the object of interpretation through the lens that the theory offers.

[Introduction]
Everyday life with ruination
Levinasian ethics
    Approach
    The Other consummated in essence
    Dislocation
    Reflexivity
    Paradoxes of landscape architecture's ethics
Meta-landscape architecture: a vision

Outline of Naama Meishar's "In Search of Meta-Landscape Architecture: The Ethical Experience and Jaffa Slope Park's Design" (Meishar 2012)

Naama Meishar's article "In Search of Meta Landscape Architecture: The Ethical Experience and Jaffa Slope Park's Design"[47] illustrates this type of outline. The article's body consists of two parts. In the first, very short section, Meishar introduces the reader to the example, the site of Jaffa Slope Park, by focusing on its complex social history. In the body's main section ("Levinasian Ethics"), the author skillfully interweaves the presentation of the theory, Emmanuel Levinas's ethics of the Other, with the presentation and interpretation of the site, giving insightful interpretations of the design process and architectural elements. The internal outline of this main section is structured around key concepts of Levinas's theory, which indicates the fruitfulness of the selected theory for making sense of Jaffa Slope Park.

One of the big pros of the theoretically informed close reading is that it forces its authors to be explicit about the chosen approach. With the exception of historians, arts and humanities scholars are usually lousy in describing their methods. But while most of their texts would gain in clarity and thus quality if they were more explicit about their method, this silence does not devaluate all scholarship—as it would if an experimental study was silent about its experimental design. This is so because the method description in scientific papers has a different function than remarks on the approach in arts and humanities papers. The necessarily subjective (which is not the same as irrational) and situated nature of the act or situation of interpretation means that any interpretation depends on and differs with the interpreter.[48] In contrast to papers in the sciences, in which the method section guarantees that a study can be repeated and that the data was correctly analyzed (e.g., that the correct statistical procedures were applied), comments regarding methods or material used in arts and humanities paper have a different function. Their main purpose is to convince the audience of the appropriateness of the interpretation. But since the ultimate proof of the pudding is in the eating (which is why Riley argues that abstracts alone do not render a good idea of a paper's quality),[49] an arts and humanities text does not stand or fall with its method description.

The Right Structure for the Job

In this essay, I have briefly sketched how the IMRaD structure has emerged in the experimental sciences to argue that it does not suit all modes of inquiry. Rather, this format caters to the needs of only one particular mode. In other disciplines, other traditions, conventions, and demands have led to the emergence of other text structures. I have presented and discussed a few of these formats, which are well-established in the arts and humanities. To conclude, I would like to argue that the aim of "raising the standards of scholarship and research, while maintaining both a diversity of approaches and some sense of relevance to real-world design"[50] is not furthered by making the IMRaD structure the default structure for academic texts within landscape architecture. This is because landscape architecture is an interdisciplinary field in a strong sense: it does not only sit between two disciplines that participate in the same mode of inquiry. Rather, it sits in between the chairs of "the three great empires of academia: the natural sciences, the social sciences, and the arts and humanities."[51] Elevating the standard structure of the sciences to the standard for the entire discipline smacks thus of authoritarianism, or even "scientific imperialism."[52] Wittingly or not, such moves inhibit thinking and writing in landscape architecture that do not participate in the scientific mode of inquiry: that is, research, or scholarship, as Riley would have us say, in the arts and humanities tradition. It would, however, be equally wrong to declare any other text structure the default for academic writing in landscape architecture. Ian Thompson has made a similar argument with regard to the use of theory. Paraphrasing him, one could say it's a matter of picking the right sort of structure for the job. He rightly emphasized that this implies that "knowing our way about becomes even more important."[53]

This insight reveals that teaching and learning of academic writing in landscape architecture are even more demanding than commonly acknowledged. Instructors and students need to be (made) aware of and at least to some degree familiar with the traditions and conventions of academic writing in *all three* empires (and of what lies in between). It is thus wrong to state that the IMRaD structure would help students understand the nature of academic writing—at best, it helps them to learn about the nature of *scientific* writing.[54] Furthermore, while it may be possible to apply the IMRaD format when writing an interpretative thesis, other variants of the essay structure will support the author in rendering the arguments in a clearer, more convincing, and more easily read fashion. Lastly, it is surely true that many landscape architecture students are overwhelmed by the prospect of writing an academic thesis. However, my colleagues' solution of offering an easy-to-follow recipe insults our students' intelligence and deprives them of the possibility to learn that creating knowledge in landscape architecture is far less rule-driven and much more creative than a one-structure-fits-all approach may have them believe. This insight may appeal particularly to students within landscape architecture insofar as it shows certain structural parallels to the design process. In any case, learning how to navigate methodological and epistemological diversity in a nondogmatic way and on knowledge-based grounds is crucial for (future) landscape architects. Modeling these dimensions of criticality may well be the most important contribution instructors can make to teaching and learning landscape architecture at a university—not in spite of but rather because of landscape architecture also being a vocational education. In a field as interdisciplinary as landscape architecture, there cannot be one default text structure. It will always be a question of choosing the right sort of structure for the job.

NOTES

1. Robert B. Riley, "Some Thoughts on Scholarship and Publication," in *The Camaro in the Pasture: Speculations on the Cultural Landscape of America* (Charlottesville: University of Virginia Press, 2015), 111.

2. Riley, "Some Thoughts on Scholarship and Publication," 111.

3. Koenraad Danneels, "'Nature's Offensive': The Sociobiological Theory and Practice of Louis Van der Swaelmen," *Journal of Landscape Architecture* 14, no. 3 (2019): 52–61, doi:10.1080/18626033.2019.1705581; Sara Jacobs, "From Novel to Relational: An Approach to Care for Relational Landscapes," *Journal of Landscape Architecture* 14, no. 3 (2019): 24–33, doi:10.1080/18626033.2019.1705571; Kevin Klosterwill, "The Shifting Position of Animals in Landscape Theory," *Landscape Journal* 38, no. 1–2 (2019): 129–46, doi:10.3368/lj.38.1-2.129.

4. Riley, "Some Thoughts on Scholarship and Publication," 119.

5. Ibid., 112.

6. John Dupré, "Against Scientific Imperialism," *PSA: Proceedings of the Biennial Meeting of the Philosophy of Science Association* 2 (1994): 374–81, doi:10.1086/psaprocbienmeetp.1994.2.192948.

7. John Swales, *Genre Analysis: English in Academic and Research Settings* (Cambridge: Cambridge University Press, 1990); Jianguo Wu, "Improving the Writing of Research Papers: IMRAD and Beyond," *Landscape Ecology* 26, no. 10 (2011): 1345–1349, doi:10.1007/s10980-011-9674-3.

8. Dwight Atkinson, "The Evolution of Medical Research Writing from 1735 to 1985: The Case of the Edinburgh Medical Journal," *Applied Linguistics* 13, no. 4 (1992), doi:10.1093/applin/13.4.337; David A. Kronick (1984), "Literature of the Life Sciences: The Historical Background," *Bulletin of the New York Academy of Medicine* 60, no. 9 (1984): 857–75; A. J. Meadows, "The Scientific Paper as an Archaeological Artefact," *Journal of Information Science* 11, no. 1 (1985), doi:10.1177/016555158501100104; Swales, *Genre Analysis: English in Academic and Research Settings.*

9. Atkinson, "The Evolution of Medical Research Writing," 340, original emphasis.

10. Charles Bazerman, in Luciana B. Sollaci and Mauricio G. Pereira, "The Introduction, Methods, Results, and Discussion (IMRAD) Structure: A Fifty-Year Survey," *Journal of the Medical Library Association* 92, no. 3 (2004): 364–67.

11. Meadows, "The Scientific Paper as an Archaeological Artefact," 27–30. Cf. Sollaci and Pereira, "The Introduction, Methods, Results, and Discussion (IMRAD) Structure." Cf. Wu, "Improving the Writing of Research Papers."

12. Wu, "Improving the Writing of Research Papers."

13. Examples of such handbooks and guides include Paul Trevorrow and Gary E. Martin, "How to Write a Research Article for MR," *Magnetic Resonance in Chemistry* 58, no. 5 (2020): 352–62, doi:10.1002/mrc.5012; P. K. Ramachandran Nair and Vimala D. Nair, "Organization of a Research Paper: The IMRAD Format," *Scientific Writing and Communication in Agriculture and Natural Resources,* ed. P. K. Ramachandran Nair and Vimala D. Nair, 13–25 (Springer International Publishing, 2014); Global Communication Center, *IMRAD cheat sheet* (n.d.), retrieved from https://www.cmu.edu/gcc/handouts-and-resources/handouts/imrd.pdf; Satu Manninen, "Academic Writing—IMRaD," *Academic Writing in English,* video (Lund, Sweden: Lund University, 2016).

14. Scott L. Montgomery, *The Chicago Guide to Communicating Science*, 2nd ed. (Chicago: University of Chicago Press, 2017), 18.

15. Atkinson, "The Evolution of Medical Research Writing," 341, 340.

16. International Committee of Medical Journal Editors (ICMJE), *Recommendations for the Conduct, Reporting, Editing, and Publication of Scholarly Work in Medical Journals*, 2019, retrieved from http://www.icmje.org/icmje-recommendations.pdf, 14.

17. Brian Hepburn and Hanne Andersen, "Scientific Method," in *The Stanford Encyclopedia of Philosophy*, ed. Edward N. Zalta (Stanford University, Metaphysics Research Lab, Summer 2021 Edition): https://plato.stanford.edu/entries/scientific-method

18. Cf. Swales, *Genre Analysis: English in Academic and Research Settings.*

19. Meadows, "The Scientific Paper as an Archaeological Artefact," 27.

20. Cf. Andersen and Hepburn, "Scientific Method."

21. Peter B. Medawar, "Is the Scientific Paper Fraudulent?," *Saturday Review*, August 1, 1964, 42.

22. Susan M. Howitt and Anna N. Wilson, "Revisiting 'Is the Scientific Paper a Fraud?': The Way Textbooks and Scientific Research Articles Are Being Used to Teach Undergraduate Students Could Convey a Misleading Image of Scientific Research," *EMBO Reports* 15, no. 5 (2014): 481–84, doi:10.1002/embr.201338302.

23. Frederic L. Holmes, "Scientific Writing and Scientific Discovery," *Isis* 78, no. 2 (1987): 220–35, doi:10.1086/354391; Karin D. Knorr Cetina, *The Manufacture of Knowledge: An Essay on the Constructivist and Contextual Nature of Science* (Oxford: Pergamon Press, 1981); Jutta Schickore, "Doing Science, Writing Science," *Philosophy of Science* 75, no. 3 (2008): 323–43, doi:10.1086/592951.

24. Schickore, "Doing Science, Writing Science."

25. Elijah Chudnoff, *A Guide to Philosophical Writing*, 2007, retrieved from https://hwpi.harvard.edu/files/hwp/files/philosophical_writing.pdf, 49.

26. Elaine Tarone, Sharon Dwyer, Susan Gillette, and Vincent Icke, "On the Use of the Passive and Active Voice in Astrophysics Journal Papers: With Extensions to Other Languages and Other Fields," *English for Specific Purposes*, 17, no. 1 (1998): 113–32, doi:10.1016/S0889-4906(97)00032-X.

27. Umberto Eco, *How to Write a Thesis*, trans. C. Mongiat Farina & G. Farina (Cambridge, MA: MIT Press, 2015), 112, original emphasis.

28. Tarone et al., "On the Use of the Passive and Active Voice in Astrophysics Journal Papers."

29. Ibid.

30. Montgomery, *The Chicago Guide to Communicating Science*, 7.

31. James D. Watson and Francis H. C. Crick, "Molecular Structure of Nucleic Acids: A Structure for Deoxyribose Nucleic Acid," *Nature* 171, no. 4356 (1953): 737–38, doi:10.1038/171737a0.

32. Montgomery, *The Chicago Guide to Communicating Science*, 22ff.

33. AWELU, "Three Versions of the RA," 2017b, retrieved from https://awelu.srv.lu.se/genres-and-text-types/writing-in-academic-genres/research-articles-ras/three-versions-of-the-ra/.

34. Elizabeth Abrams, "Essay Structure," 2000, retrieved from https://writingcenter.fas.harvard.edu/pages

/essay-structure; Wayne C. Booth, Gregory G. Colomb, Joseph M. Williams, Joseph Bizup, and William T. Fitzgerald, in *The Craft of Research*, 4th ed. (Chicago: University of Chicago Press, 2016), 177; Eco, *How to Write a Thesis*, 107.

35. Abrams, "Essay Structure."

36. Eco, *How to Write a Thesis.*

37. Booth et al., *The Craft of Research;* Abrams, "Essay Structure"; AWELU, "Three Versions of the RA."

38. Eco, *How to Write a Thesis*, 112.

39. Rebecca Wingfield, Sarah Carter, Elena Marx, and Phyllis Thompson, *A Guide to Researching and Writing a Senior Thesis in Studies of Women, Gender, and Sexuality* (o.a.), retrieved from https://hwpi.harvard.edu/files/hwp/files/womens_studies_senior_thesis.pdf?m=1370451715, 57.

40. Kenny Cupers, "Gardening as Geopolitics," *Journal of Landscape Architecture* 14, no. 3 (2019): 46–51, doi:10.1080/18626033.2019.1705580.

41. K. Walk, "How to Write a Comparative Analysis," 1998, retrieved from https://writingcenter.fas.harvard.edu/pages/how-write-comparative-analysis.

42. James Corner, "A Discourse on Theory II: Three Tyrannies of Contemporary Theory and the Alternative of Hermeneutics," *Landscape Journal* 10, no. 2 (1991): 115–33, doi:10.3368/lj.10.2.115.

43. Simon Rippon, "A Brief Guide to Writing the Philosophy Paper," 2008, retrieved from https://writingproject.fas.harvard.edu/files/hwp/files/bg_writing_philosophy.pdf?m=1370481947.

44. Susan Herrington, "Framed Again: The Picturesque Aesthetics of Contemporary Landscapes," *Landscape Journal* 25, no. 1 (2006): 22–37.

45. Herrington, "Framed Again," 26.

46. Michelle Syba, "A Brief Guide to Writing the English Paper," 2008, retrieved from https://hwpi.harvard.edu/files/hwp/files/bg_writing_english.pdf.

47. Naaman Meishar, "In Search of Meta-Landscape Architecture: The Ethical Experience and Jaffa Slope Park's Design," *Journal of Landscape Architecture* 7, no. 2 (2012): 40–45, doi:10.1080/18626033.2012.746086.

48. Hans-Georg Gadamer, *Truth and Method*, 2nd ed., trans. William Glen-Doepel, revised by J. Weinsheimer and D. G. Marshall (New York: Crossroad, 1989). For a concise introduction to Gadamer's thinking, see Jeff Malpas, "Hans-Georg Gadamer." *Stanford Encyclopedia of Philosophy*, ed. Edward N. Zalta, Fall 2018, https://plato.stanford.edu/entries/Hans-Georg Gadamer.

49. Riley, "Some Thoughts on Scholarship and Publication," 112.

50. Ibid., 111.

51. Ian H. Thompson, chap. 3 in *Research in Landscape Architecture: Methods and Methodology*, ed. Adri van der Brink et al. (Abingdon, Oxon–New York: Routledge, 2017), 40.

52. Dupré, "Against Scientific Imperialism."

53. Thompson, chap. 3 in *Research in Landscape Architecture*, 49.

54. Cf. Susan M. Howitt and Anna N. Wilson, "Revisiting 'Is the Scientific Paper a Fraud?'" 481–84.

More Important Questions

M. ELEN DEMING

Prologue

I met Bob Riley while attending my first CELA (Council of Educators in Landscape Architecture) conference at the University of Oregon in fall 1993. A decade or so later our paths began to run parallel, then bend, intersect, and overlap for a short time. Eventually I would serve as co-editor of *Landscape Journal* (for which Bob remained consulting editor) and, in 2008, I joined the University of Illinois as professor and head of the Department of Landscape Architecture. At Illinois, Bob became a supportive mentor, trusted colleague, and, gradually, a treasured friend. During his last decade, plagued by health issues, loneliness, regret, and nameless other challenges, he wrestled with his legacy. Against that backdrop, producing his volume of selected essays—*The Camaro in the Pasture* (2015)—required remarkable faith and determination. I still read the title as a poignant metaphor for the way Bob saw himself in his later years. Through it all, to be graced by his honesty, charm, warmth, and intellectual generosity was a gift beyond measure.

On Questions

Robert Riley was an eclectic, puckish sort of savant who did not suffer fools. His unique voice as a critical essayist also made him a thought leader for a generation of thinkers and writers in landscape architecture. In particular, this essay acknowledges Riley's contributions as editor of *Landscape Journal* and provocateur during a transformative period in the field.

During Riley's editorship (1987–95), long before the proliferation of social media platforms, blogs, and open-access journals we enjoy today, *Landscape Journal* was the first and only peer-reviewed journal in the field, the communications instrument of CELA. Certainly there were important journals in cognate fields (ecology, geography, land planning, architecture, etc.). During that period, aside from a few other, school-run publications—for example, Louisiana State University's *Critiques of Built Works of Landscape Architecture* (1994–ca. 2005); *Princeton Journal: Thematic Studies in Architecture* (notably the *Landscape issue* of 1985); Harvard's *GSD Magazine* (currently in its forty-eighth issue); *Penn in Ink* (now *PennDesign Annual*), and so on—opportunities for public debate about landscape architecture typically were limited to annual conferences.

In his first editorial as sole editor of *Landscape Journal,* Riley wryly observed, "practitioners have magazines and scholars have journals." He acknowledged a set of hybrid publications (e.g., *Society and Space; Places,* and others) evoking timely, spirited discourse on the kind of topics that rarely appeared in the mainstream mouthpieces and added: "Much of our best philosophy, criticism, and speculation come from these 'in between' publications."[1] It signaled his editorial agenda—neither stodgy academic reporting nor a blunt buttress to conventional practice, but something more fluid, intellectually open-minded, provocative, and, above all, rigorous.

Certainly, Riley was not afraid to use his editorial platform to provoke a good discussion. Together with assistant editor Brenda Brown, his intellectual partner at the *Journal* (from 1990 to 1995), he prodded readers to examine their assumptions, standards, experiences, and responsibilities, and experimented with a series of editorial gambits. The remarkable essays they received were published as discussion papers: "Most Influential Books" (1991), "Most Important Questions" (1992), and "Most Influential Landscapes" (1993). These pieces in turn joined other multi-authored discussions, framed and introduced by leading lights, such as Margaret McAvin's "Landscape Architecture and Critical Inquiry" (1991).[2] Regular issues of the *Journal* were augmented by guest-edited theme issues such as *Women, Land, Design* (1994).[3]

Concurrently, a series of other topical provocations initiated (though perhaps *fomented* is more accurate) by Riley and Brown rippled right through the 1990s. There were substantive conversations on the limits of intended form and meaning, as well as on nativism in ecological design. Notably, as a follow-through from the *History in Landscape Architecture* symposium held at Penn State (June 1994), Riley launched his 1995 polemic "What History

Should We Teach and Why?" Responses followed from Catherine Ward Thompson and Peter Aspinall's "Making the Past Present in the Future: The Design Process as Applied History" (1996)[4] and an alternative view from Dianne Harris: "What History Should We Teach and Why? An Historian's Response" (1997).[5] The series remains a thoughtful springboard for teaching history.

In short, Riley's editorship of *Landscape Journal* "raised the bar" for the scholarly community in landscape architecture for a generation. While not all published in *Landscape Journal,* such extended, thoughtful debates served as provocation for a surge of scholarship to come in the late 1990s and early 2000s. During these formative years, in advocating for topics of research that seriously addressed consequences for professional design, Riley and Brown moved the whole profession forward.

OUR OWN QUESTIONS

As a young student of philosophy (University of Chicago), Riley would have learned that it is the relationship of worldview to intellectual curiosity that affords an ability, indeed the *privilege* to ask disciplined questions. Inquiry is a social construction after all. Our education may blinker or liberate our understanding of what is possible or even permissible to ask. Questions seem to run in packs; though given the critical importance of learning research skills in professional programs, perhaps they 'swim in schools."[6]

Indeed, throughout Riley's long career, the rigor of inquiry itself served as connective tissue. His fascination, sharply lit in the 1992 discussion "Most Important Questions," offers us vantage on the arc of landscape architectural thinking that encompasses his editorship (1987–95). Just before entering that period, in introducing the 1985 Illinois CELA conference proceedings, Riley writes: "Our maturity as an academic discipline, in distinction from a profession, lies not in journals nor doctoral programs *per se,* but in forming and asking *our own questions:* questions amenable to research or scholarship, to interpretation or speculation, but questions aimed at producing answers that will make us better shapers of the human habitat."[7]

Later, in "Gender, Landscape, Culture: Sorting Out Some Questions" (*Women, Land, Design* issue), Riley remarked: "Consideration of important questions within our discipline, and within feminist inquiry as a whole, should precede questions about gender and landscape. . . . The author concludes with a plea for more rigorous inquiry."[8]

The editorial gloves came off completely in "Some Thoughts on Scholarship and Publication" (1990), where Riley registers a trenchant critique of the poor research standards inculcated in professional programs at the time: "It is certainly clear that the academic design community now communicates largely with itself, has established its own set of awards and peer recognition, socializes its children, provides rites of passage, and above all poses its own questions and develops its own paradigms distinct from, and sometimes irrelevant to, the world of practice."[9] In its evolution from a profession to a discipline, he observed, a lamentable gulf now yawned between the subcultures of academia and practice in landscape architecture, "a subject for study in itself as an exercise in the sociology of disciplines."[10] But in response to concerns that the *Journal* might somehow be 'dominated' by an empirico-quantitative cabal, Riley threw down the editorial gauntlet in defense of rigor:

> People who submit quantitative-analytical manuscripts to *Landscape Journal* have usually received doctoral training in their specialized field . . . [or] have learned to do research. Unfortunately, many of the people who submit "think pieces" have never learned to think. Anyone who takes offense at this harsh characterization should be forced to read some of the drivel submitted under the rubric of design philosophy or theory. . . . The hard fact is that the "humanist" or speculative tradition in landscape architectural scholarship will hold its own with quantitative research in this journal when, and only when, academic landscape architecture programs start teaching their students—and their faculty—the difference between sloppy concern and rigorous thought.[11]

It may be helpful to put Riley's editorial provocations in the context of his time. Between the mid-1980s and the close of the millennium, we see a sharp rise of activity in design theory and speculative work. Part of the impetus was a response to adjacent fields, for instance sculpture and architecture. But over time, after the late 1970s recession and early 1980s neoliberalism (e.g., "trickle-down economics"), reductions in public funding meant growing institutional competition. Recruiting and retaining distinguished research and thought leaders (a leading indicator of academic success), it was hoped, might anchor talented faculty and thus the reputation (the lagging indicator of academic success) of professional schools.[12]

Claiming *our own* expertise, building *our own* body of knowledge, expanding the methods, and clarifying the standards for knowledge production remain crucial ways for scholars to influence the scope and mission of a profession. Yet the art of effective inquiry requires depth of training, discipline, practice. It also requires confidence. As written elsewhere, "better questioning helps produce better answers—and therefore new competencies."[13] In landscape architecture in particular, scale-shifting, translating, intersecting, and reframing are constantly required to maintain relevance in response to contemporary need—relevance being a key metric.

Research questions are the golden currency of any discipline. The best expression of core disciplinary mission and values may be seen in the range and scale of questions it addresses. Questions are (or should be) egalitarian, alternately leveling and opening the field of inquiry; everyone can (and should) ask good questions, especially students, clients, citizens—even designers.[14] This is so because in the reciprocity of design knowledge, both in its production and consumption, "new knowledge is never a one-way street, and expertise does not naturally flow from the academy to the profession."[15]

Most Important Questions

"Most Important Questions" started as a follow-up to discussions begun at the 1991 CELA conference, an open-ended lob to a score of leading influencers in the field of landscape architecture.[16] Riley and Brown challenged twenty respondents to identify the fundament(s) and firmament(s) of landscape architecture upon which all manner of scientific, humanistic, or aesthetic, thoughts might be scaffolded. "What do you consider the most important question(s) in landscape architecture today?" they wrote: "We particularly have in mind questions that suggest directions for continuing inquiry, questions that seek conceptual and intellectual advances as well as practical ones. Please explain why your question is so important. You might also address whether or not the question is answerable, who might answer it through what methods (from 'hard research' through speculative musings), and what difference this question and its answer(s) might make."[17] In 1991–92, the recruits comprised a set of contemporary elites who (the editors felt) were: (1) actively shaping the discourse of the field and (2) capable of preparing thoughtful and fairly well-developed responses. This demanded both editorial access and trust (though chits undoubtedly were called in). And in what other universe could an editor pull this off—but for a fairly exclusive club in a very tiny field?

The resulting discussion paper remains a treasure trove containing insights of depth and power. By any estimation, it remains a fascinating historical snapshot of the landscape intelligentsia in the early 1990s. Upon rereading, I am suffused by respect, affection, gratitude for the diligence and earnest naiveté of the respondents. As contributors to the intellectual history of the field, every one of them worked hard to address the borders and capacities of landscape architecture as a body of knowledge.

Describing the responses as "mind-numbing in their range and diversity," Riley and Brown avoided curation beyond a few brief summary observations. "Many questions are rhetorical, posed in preambles to position statements. Many deal with what we need to be or do, not with what we need to find out. Many are concerned with structuring inquiry, information, knowledge, and understanding."[18] Nevertheless, the discussion holds a mirror to the still-emerging discipline.

INQUIRY AND STRATEGY

What can "Most Important Questions" tell us about the distance the field has traveled since 1992? Riley noticed a tendency to focus less on any specific "need to know" in favor of questions about who to serve, what values to uphold, and what outcomes to secure. For instance, Setha Low (Cultural Anthropology, Graduate School and University Center of the City University of New York) invokes "three inseparable kinds of intellectual inquiry: (1) ecological, (2) political, and (3) sociocultural," which together "generate the conceptual questions that guide research and practice."[19] Stephen and Rachel Kaplan (Environmental Psychology, University of Michigan) philosophically ask: "[I]s nature good [for people]?" And if so, how? How essentially? As an amenity, or in some other way? And finally, they ask, "which nature is it that does all these good things?"[20]

These kinds of global inquiry maintain some contemporary resonance. But for Randolph Hester (Landscape Architecture and Environmental Planning, University of California, Berkeley), whose responses strike one with an uncanny resonance today, there are many other things we simply could not have heard expressed in 1992:[21] for example, very few questions centered on social or environmental justice, racial inequality, poverty, migration, or accessibility. Neither climate change nor pandemics were on the radar yet; sea level rise and climate refugees were not a "thing." The world was not yet as apocalyptic as it sometimes seems now to be.

Yet within individual responses there remains much to observe, both on the surface and below. We might begin by observing *framing strategy.* Some rejected the premise outright, before obediently plowing into a manifesto. A few present a single umbrella question with a cluster of genetically linked questions clustered round its skirts. Many laboriously compile a comprehensive laundry list of topics that span the waterfront. Still others articulate a detailed research agenda.

There is the *strategy of alliance.* In 1991–92, landscape architecture was busy establishing its disciplinary identity by bracketing its knowledge. To detach from long formal and programmatic attachment to (and intellectual dependence on) architecture, it needed to demonstrate *gravitas* of its own. Many scholars professed alternative alignments with the critical aims of cultural geography and/or the bioethics of landscape ecology. Such fields were already intellectually well-developed but, lacking the pragmatic applications of the design profession, were ripe for transmutation to the site scale.

Perhaps we could examine *intellectual paradigms:* many respondents avoided questions of "what we know" (ontology) in favor of "how we know it" (epistemology). Or as Riley put it, "Many deal with what we need to be or do, not with what we need to find out." In other words, what guides the scope and mandate of the field when there seems to be no consensus on either the questions or, indeed, the knowledge paradigm(s) most appropriate for landscape architecture?

Toward *investigative strategy* and the crafting of researchable questions, Richard Chenoweth (University of Wisconsin) comments on the crucial importance of precision. If the DNA of every answer is held within the seeds of the originating question, then we must attend carefully to "the form that questions should take":

> Failure to devote considerable energy to this issue results in publications full of answers, that is, alleged truths, to questions that were never posed in the first place. . . . Poorly constructed questions, if answerable at all, fail to guide the inquiry in a manner that the investigator either gathers the relevant facts or is able to rule out explanations, other than those embraced by the investigator, for the facts as observed. At the extreme, investigators venture forth without knowing whether their purpose is to describe phenomena, predict

outcomes, or determine cause and effect relationships. Because each purpose requires a different modus operandi, not knowing one's purpose is likely to result in empirical trivia, theoretical bullshit, or both.[22]

Perhaps so many different questions are suggested because so many "landscape architectures" are in evidence. The sheer breadth of inquiry may serve as a point of analysis: ranging from objectivist or normative agendas (environment and behavior studies guiding policy implications; design guidelines, etc.) to constructivist or interpretive agendas (landscape narrative and affect; form and poetics; belonging and attachment; reciprocity between humanity and nature' etc.) and critical or pragmatist agendas (research through design; participatory design for social impact; critical pedagogy; etc.). Given the impossibility of identifying a central disciplinary research problem, let us at least appreciate the array of worldviews on display. It is also noteworthy that the thought leaders of this period engendered design pedagogies so fertile that they in turn shaped distinct identities—schools of thought, if you will—for their respective institutions.[23]

OBJECTIVISM/POSITIVISM

Objectivists lay out research agendas that are generally predictive, instrumental, and/or solution-oriented. Objectivist questions tend to description, measurement, modeling, and controlled treatment. The gold standard of positivist inquiry is to test theory using necessarily reductive methods in order to eliminate potentially confounding variables, concentrate possible explanations for events, forestall the vagaries of uncertain outcomes, and enhance return on treatment/investment. Given the complexity of landscape and human interactions, though, it is fairly rare to find strict adherents to this mode in landscape architecture.

The expansive questions set out by Clare Cooper Marcus (Architecture and Landscape Architecture, UC–Berkeley) demand objectivist research designs using observation, description, or direct measurement.[24] Others in this vein include Richard Chenoweth (Environmental Psychology, University of Wisconsin) on problems of visual quality, a dominant research agenda from the mid-1970s to the early 1990s. The instrumentality of studying landscape aesthetics appears where public reception of infrastructure projects might be managed, economic benefits of "pleasing environments" might accrue (to public investors), and the values of environmental decision makers might be guided.[25]

CONSTRUCTIVISM/INTERPRETATION

Constructivism assumes that all knowledge is socially constructed and mutable, as well as meaningful to the community that produces it. Thus, constructivist inquiry seeks to discern and interpret codes or patterns in situ, or contextualized in practice. Codes may be used for spatial organization, for telling stories, or developing normative guidelines and principles. Constructivists question issues of values, perception, attachment, and interpretive meaning.

Articulating their national research agenda in the early 1980s, Jim Palmer (professor *emeritus* of landscape architecture and visual assessment, SUNY College of Environmental Science and Forestry [ESF]) and Rick Smardon (distinguished professor *emeritus* of environmental planning, SUNY ESF) used social surveys to construct four overarching themes: landscape management and stewardship; social meaning of landscape; design for the "new age" (sustainability); and ethical bases for the profession.[26] Arnold Alanen (Landscape Architecture, University of Wisconsin, and founding editor of *Landscape Journal*) and Evelyn Howell (Plant Ecology, University of Wisconsin) address core questions of history, theory, and criticism, asking "How do landscapes change?"[27] And by declaring that "language has consequences. It structures how we think and what kinds of things we are able to express," Anne Spirn (Landscape Architecture, University of Pennsylvania) lays out her own gorgeous constructivist understanding of codes of meaning.[28]

PRAGMATISM/ACTIVISM

Pragmatist questions inform design processes, strategies, and outcomes, where designers and engaged activists may strike for social impact and emancipation. "How?" questions are addressed through pragmatic action and engaged research, including critical pedagogy, design prototype, and proof-of-concept studies. They may be speculative, synthetic, or critical, but always tend to expand design possibilities. Questions cluster around making (rather than meaning) processes, such as what type of design operations may intensify or multiply the promise of the field.

In observing the "fluctuating and permeable" boundaries of the landscape discipline, Margaret McAvin (Landscape Architecture, Rhode Island School of Design) suggests critically pragmatic investigation through the work of design itself: "Landscape architecture's primary mode of scholarly inquiry is the creation of landscapes."[29] James Corner (Landscape Architecture, University of Pennsylvania) echoes McAvin, pronouncing that design itself may be a pathway for new knowledge. Invoking pragmatist guru Richard Rorty (1931–2007), Corner recognizes new knowledge in design as transactional: "[In] a time where, to quote Rorty, 'everything is a conversation,' then the art of asking and responding to questions is something we need 'to cultivate and bring to blossom.'"[30]

More Important Questions

Robert Riley was our crusty muse during a moment of enormous intellectual expansion in landscape architecture. Brenda Brown's "*Tribute*," while describing Riley's influence as a department head, also implies his wider impact on the field: "Colleagues from that time characterize his leadership as intellectual . . . in exercising an inquisitiveness and ability to understand, articulate, support and show confidence in the work and worth of people whose interests were very different from his own."[31] Riley demanded deeper intellectual rigor and diversity of perspective from his entire community. And we are grateful for that.

IMPERATIVES

Fast forward. Landscape architects today are showing greater confidence, exercising fierce vision, skill, drive, and influence. For heaven's sake, two landscape architects have won MacArthur "Genius" grants in the last two years alone. So how, and how fast, are we shaping "our own questions" to meet the imperatives of our time? Have the goals of our practices changed? Or are our tools and methods simply changing faster than our values?

Any response demands perspective on disruptions, transformations, *and* continuities in the field. Following the attacks of September 11, 2001, after the general shock, a new chapter opened for landscape architecture along with many other professions and institutions. New design projects involved security measures for public infrastructure; industrial and adaptive reuse; vacant-land ecosystems services; multicultural communities. Research demanded new methods and tools for sustainability and landscape performance metrics; evidence-based design; carbon consumption and offsets; climate risk assessment; engaged and transdisciplinary action research.

Cultural changes have only accelerated since the 2008 recession and recovery. Today's important questions increasingly run across and between industry sectors, in hybrid practices and teams connecting practitioners with faculty specialists. Designers are irresistibly caught up with media, digital technology, and experience design. The "internet of things" is also the "internet of places" as urban spaces are increasingly networked to send and receive surveillance and geo-tagged data. All this will undoubtedly continue to affect the conduct of the field, seen by many (with cause) to be uncritically complicit in the dominant Western culture of structural racism and ruthless greed.

ESSENTIAL VOICES

Who should represent and expand thought leadership today? Compared to 1992, surely they should include more people of color; surely more women. As the field has feminized over the past quarter-century, scholarly leadership mirrors that trend (although professional design leadership is still catching up). But our respondents need to be diversified in many other ways as well: culturally (rural/urban/religious/not), racially (Black/Brown/Indigenous), demographically (age/income/education), and politically (conservative as well as liberal).

And while it has been instructive to include voices from adjacent fields (e.g., architecture, planning, and ecology), how about including those in K-12 education, public medicine, user-centered and computational design, machine learning and human-computer interactions, or social entrepreneurship? We need only look at the Landscape Architecture Foundation's professional development programs (e.g., Deb Mitchell Research Grant; Fellowship for Innovation and Leadership; Olmsted Scholars Program) to see a future where greater diversity of all kinds is strengthening the field.[32]

With a differently conceived, more inclusive group, surely we would hear a good deal more on landscape architecture as political advocacy or emancipatory activism, ideas with deep historical roots in the field. Many would promote the values of public health (consider why landscape architecture first emerged in nineteenth-century cities) as a principal driver of research questions. The pandemic has torn open the economic and spatial risks inherent in the status quo and reminded us all (surprise!) that public open space can help save lives.

FOUR ISSUES

In reflecting on Riley's provocative "Most Important Questions," we would invite discussion of even *more important questions* facing the field today. Let me venture my own four issues. Each evokes a rather dark agenda; admittedly the associated research questions remain inchoate. Further, the implied questions are necessarily transdisciplinary, because none can be limited to the medium of landscape.

1. Limits of ideology. The capitalist distribution of goods and services has resulted in intolerable social inequities and obscene wage gaps. In the Darwinian wreckage of the social contract, the social landscape is breaking too, with bias and fear wielded as political levers of alienation. Antiracist landscape restoration may be our best chance for landscape justice extended to everyone. What kinds of advocacy and engagement can landscape architects undertake to right such massive wrongs?

2. Limits of growth. A phrase that once referred abstractly to planetary carrying capacity and human population control now encompasses spatial competition for settlements, water, food, nonhuman habitat, and resource extraction. Inherent in the myths of perpetual growth and unlimited ecosystems services there is vast collateral damage—the "Tragedy of the Commons" run amok in habitat encroachment, deforestation, and eco-disruption at a continental scale.[33] We have passed the tipping point of climate instability. Acceptance of resource scarcity and the true cradle-to-cradle costs of harvest and waste cycles demands new approaches to land use; adaptive farming and water harvesting; urban nature and rural production. What innovations can landscape architects devise for enhancing landscape resilience?

3. Limits of utopia. The unintended consequences of artificial intelligence and other technological fictions will continue to exacerbate climate change and inequities. As humans overrun the planet (8.8 billion and counting fast), and environmental systems cease to function, becoming increasingly inhospitable, we supplant it with virtual landscape experiences, gaming, models, and simulacra. In dystopia, how might landscape architects contribute to cognitive health and decision-making through the design of digital, cyborgian, and synthetic landscapes?

4. Limits of landscape architecture. Is the discipline up to the tasks ahead? That may depend on how you define the field—the notion that landscape architecture is distinct and autonomous from other fields may already be outmoded. Simply applying critical

design skills to the physical medium of landscape, over temporal scales of seasons and cycles, with spatial scales ranging from micro-sites to watersheds and regions, may not be enough. We are too small a field to change the world alone, and so must multiply and amplify our voices politically as well as enlist ideas and engage designers from as many other fields as possible. How will landscape architects leverage their skills as advocates to persuade others to become planetary landscape actors?

Conclusion

As in every other field, landscape architecture faces threats and risks, disruption and division: in the ubiquity of social media lie the potential manipulation of popular opinion; growing distrust of professional expertise; targeted conspiracy and misinformation; splintering of the social contract and political nihilism; and the surfacing of racism and white supremacy after festering for so long underground. In a near-constant state of damage control, we sense the end of the utopian ideology that has been a core driver of the landscape architecture project from the beginning.

But there are opportunities and encouragements too: long strides have been taken to embrace diversity, inclusion, and equity; growing awareness of antiracist and anticolonial design; more and better subaltern histories; general recognition of environmental injustice in the face of global climate change; better access to design tools and methods; and DIY design, participatory action, and community can-do spirit. The healing benefits of near nature still work their magic upon our brains and bodies. And steady improvements in practice-based design research mean that our collective efforts to build knowledge are being multiplied and applied.

Riley never predicated landscape architecture's "maturity as an academic discipline" on intellectual autonomy: hermeticism makes no sense for a field built on expansive thought and free-range transdisciplinarity. And while we are at it, let's acknowledge too that learning on the fly is a formidable skill set, essential for professional practice, and imparted very well by many educational programs. But whatever logistical revelations may be presented by big data or computational design, it remains for landscape architects to define the salient and research(able) questions to be found there. Pressing, emerging problems demand a disciplined approach, lest we lose our already friable sense of coherence—not to mention ethics.

As someone once said, we tend to measure what we value—with the corollary that we value what we can measure.[34] The profession has increasingly embraced performance-driven outcomes for design (versus, say, meaning-driven). Yet if we are honest, we are forced to admit that many categories of landscape performance are not measurable, at least not in terms of conventional research methods. One thinks of Beth Meyer's essay "Sustaining Beauty: The Performance of Appearance" (2008),[35] for example, to understand that landscape affect, a sense of the uncanny, or personal self-knowledge, may not yield easily measurable outcomes. But why should that mean they are not valuable?

The maturation of landscape architecture as a body of specialized knowledge demands that we continue to clarify what we value and find new ways to sense, recognize, and measure it—measuring how well landscape *performs* more diverse values. It may be helpful to think of disciplinary research as an organism—an owl, for instance, with its organ systems articulated into distinct parts each serving the growth of a whole being, a whole body. Depending then on the organ system under study, the most important questions that may be asked, along with the methods, data, and analysis needed to generate answers, will need to be as functionally and poetically discrete as wings and beaks, eyes and talons.

The ability to craft good questions, many more important questions, has never been more critical. As Margaret McAvin put it: "A discipline is only as important to its culture as the questions it asks, and as significant as the answers it generates."[36] What is promising is that in its ability to learn quickly and re-tool, landscape architecture shows itself to be intellectually resilient, strategically pliable, and intentional in response to the extraordinary questions of Now. Because the field of landscape architecture is, and never has been, just one thing: it is manifold, intellectually complex, functionally and perceptually mutable.

Very much like our friend, Robert Riley.

NOTES

1. Robert B. Riley, ed., "Editorial Commentary: Some Thoughts on Scholarship and Publication," *Landscape Journal* 9, no. 1 (Spring 1990): 48.

2. Margaret McAvin, ed., "Landscape Architecture and Critical Inquiry," with Elizabeth Meyer, James Corner, Hamid Shirvani, Kenneth Helphand, and respondents Robert Riley, and Robert Scarfo, *Landscape Journal* 10, no. 2 (Fall 1991): 155–72.

3. John Furlong and Karen Madsen, *Landscape Journal* (Women, Land, Design) 13, no. 2 (Fall 1994).

4. Catherine Ward Thompson and Peter Aspinall, "Making the Past Present in the Future: The Design Process as Applied History," *Landscape Journal* 15, no. 1 (Spring 1996): 36–47.

5. Dianne Harris, "What History Should We Teach and Why? An Historian's Response," *Landscape Journal* 16, no. 2 (Fall 1997): 191–96.

6. In some exasperation, Riley once asked: "If we don't know the answers to any of these questions (and we don't), and if we don't discuss them openly (and we don't), how in the world are we teaching students or evaluating their work?" In Robert B. Riley and Brenda J. Brown, eds., "Most Important Questions," *Landscape Journal* 11, no. 2 (Fall 1992): 181.

7. Robert B. Riley, "Foreword," *Proceedings of the 1985 CELA Conference: Prospect, Retrospect, Continuity*, ed. Brian Orland (Champaign, IL: CELA, 1985), iii. Emphasis added.

8. Robert B. Riley, "Gender, Landscape, Culture: Sorting Out Some Questions," *Landscape Journal* 13, no. 2 (Fall 1994): 153, doi:10.3368/lj.13.2.1531994.

9. Riley, "Editorial Commentary," 48.

10. Ibid., 48.

11. Ibid., 50.

12. Ranking industries ballooned thereafter, e.g., *Design Intelligence*, founded in 1994. https://www.di.net/sponsored-research/

13. M. Elen Deming, "Editor's Introduction," *Landscape Journal* 28, no. 2 (Fall 2009): vi.

14. Though contested by some, a belief that practitioners may be capable of producing high quality research has taken aspirational root in many design firms. Evidence that practitioners are increasingly engaged in knowledge transactions may be seen in annual ASLA and CSLA awards programs, as well as the number of hybrid research teams that include designers in prominent roles. One need only cite the Landscape Architecture Foundation (LAF) *Summit on Landscape Architecture and the Future* (Philadelphia, June 2016) or the Royal Institute of British Architects (RIBA) Annual Research Awards for dozens of useful exemplars.

15. Deming, "Editor's Introduction," vi.

16. In a personal communication, Brenda Brown points out that although this statement "might be taken to mean that the respondents were all landscape architects . . . of course they weren't. The inclusion of Setha Low and the Kaplans, even Clare Cooper Marcus, for example, reflects Bob's general interdisciplinary bent as well as his work with the Environmental Design Research Association (EDRA)." October 12, 2020.

17. Riley and Brown, "Most Important Questions," 160.

18. Ibid., 160.

19. Ibid., 172.

20. Ibid., 162. Stephen Kaplan (1936–2018) died recently.

21. Although ostensibly rejecting the entire premise, in 1992 the singular Randolph Hester offers a research framework that has weathered handsomely over the last quarter-century. "Central questions change over time," he writes, "not as fast as fashion designers wish . . . but not as slowly as geological time" (161). He then lays out a set of seven research agendas: environmental anomie ("our accelerating disassociation from the world"); landscape

justice ("just public landscapes"); new democracies (bottom-up design, innovation, and community); revitalizing environmental agencies and regulations; the resilient city region (new parameters of 'sustainability'); the political landscape ("holistic landscape decisions"); and education in Oz ("how do we . . . continue to educate specialized generalists?" (161). It strikes me as remarkable that every element in Hester's outline rings just as true today.

22. Richard Chenoweth, in Riley and Brown, "Most Important Questions," 170.

23. For example, after Ian McHarg's retirement, Anne Spirn, James Corner, and John Dixon Hunt reintroduced phenomenology and critical studies in cultural landscape at the University of Pennsylvania; Reuben Rainey, Warren Byrd, and Beth Meyer developed the contemporary emphasis on sensitive site analysis and interpretive design at the University of Virginia; Stephen and Rachel Kaplan made the University of Michigan's reputation as a scholarly powerhouse in environment and behavior research; and at SUNY's College of Environmental Science and Forestry, Jim Palmer and Rick Smardon helped define expertise in visual assessment for environmental management and policy. There are other potent examples at University of California–Berkeley, University of Oregon, University of Georgia, and elsewhere.

24. Clare Cooper Marcus, in Riley and Brown, "Most Important Questions," 174–76.

25. Chenoweth, ibid., 170; Cooper-Marcus, Chenoweth, Kaplan, and Hester are professors emeriti today.

26. James F. Palmer and Richard C. Smardon, in Riley and Brown, "Most Important Questions," 178. Arguing for an enhanced research imperative in practice, Palmer and Smardon emphasize landscape performance, an idea that has now taken firm root in the Landscape Architecture Foundation's Landscape Performance series (https://www.landscapeperformance.org/), as well as Landscape Architecture Accreditation Board standards.

27. Arnold Alanen and Evelyn Howell in Riley and Brown, "Most Important Questions," 169. Alanen and Howell point to the paucity of subaltern landscape histories, adding that "the maturation of landscape architecture as a discipline will be revealed not only by those elements of the past and present that are incorporated into the historical corpus" (169–70). It is somewhat satisfying to observe how much work has been advanced in the past twenty-five years (though so much is still needed).

28. Anne Whiston Spirn, in Riley and Brown, "Most Important Questions," 180. Alanen, Palmer, and Smardon are professors emeriti today. Spirn is Distinguished Professor of Landscape Architecture and Planning at MIT.

29. Margaret McAvin in Riley and Brown, "Most Important Questions," 173. Margaret McAvin (1946–1999) died a few years after her essay was published.

30. Richard Rorty, *Philosophy and the Mirror of Nature* (Princeton, NJ: Princeton University Press, 1979), cited by James Corner in Riley and Brown, "Most Important Questions," 164. In his essay, Corner launches a question that has dramatically reshaped practice and pedagogy since 1992: "How might landscape architecture participate in, if not actually lead, a critical globalism?" (163–64).

31. Brenda J. Brown, "A Tribute to Robert B. Riley, 1931–2019," *Landscape Journal* 38, no. 1–2 (January 2019): 2, doi:10.3368/lj.38.1-2.

32. Landscape Architecture Foundation, "Scholarships," Washington, DC, https://www.lafoundation.org/what-we-do/scholarships.

33. Garret Hardin, "The Tragedy of the Commons," *Science* 162, no. 3859 (1968): 1243–1248, doi:10.1126/science.162.3859.1243.

34. This truism has taken root in the business world particularly. See Ariely, "You Are What You Measure," *Harvard Business Review,* June 2010, https://hbr.org/2010/06/column-you-are-what-you-measure.

35. Elizabeth Meyer, "Sustaining Beauty: The Performance of Appearance. A Manifesto in Three Parts." *Journal of Landscape Architecture* (Spring 2008): 6–23.

36. McAvin, in Riley and Brown, "Most Important Questions," 174.

Two views of the garden at Ryoanji.
(Photographs by Robert B. Riley)

II. Landscape History

TEACHING AND PRACTICE

1995

What History Should We Teach and Why?

ROBERT B. RILEY

Landscape Architecture faculty seem increasingly concerned with the role of history in design and in the curriculum. Part of this interest is probably a cycle of fashion: a reaction against the modern movement's abandonment or debasement of history. Part is increasing intellectual sophistication amongst our landscape architecture departments, a dissatisfaction with sabbatical slides serving as the substance of history. Part might be spillover from the revival of historical forms in postmodern architecture. No small part of it is the pressure upon landscape architectural faculty to produce something "scholarly," which often means amateurish history *cum* cultural geography.

This renewed interest raises a number of questions, some of which have been around for a long time. Is one undergraduate history course enough? If not, what should the following courses be, and what is their relationship to one another? Should "the" history course continue to follow the five-hundred-years-of-Western-Civilization-a-month model? And what is a history course? The three courses I teach all deal with the landscape in different periods, and major parts are organized chronologically, but this does not make them history courses. Should history be devoted to learning the details and craft of a profession, or should it be a general education elective for all units? (Too often the answer is both.) Is the standard introduction to landscape architecture course basically a history course with a few token visiting practitioners? Should it be? If the history course is to seriously inform our students' design, is it best given at the beginning of their studio track, or is it more provocative and useful at the end? These are simple and obvious questions, familiar to all of us. There are, however, some more basic issues of the *what* and *why* of history.

Our Distinctive Use of History

History is a microcosm of our profession and our discipline. For the last quarter of a century, many things have been added to it and almost nothing deleted. As an example, consider what the following terms have in common: *Paleolithic Period, pastoralism, patio homes, peasants, pesticides, plank roads, post modernism, pre-Colombian landscape*, and *proto-literate culture*. Yes, they all start with p and seem vaguely related to landscape history. Beyond that? They are all terms found in the subject index of Philip Pregill and Nancy Volkman's *Landscapes in History* (1993) but not in Norman Newton's *Design on the Land* (1971). Our additions to history reflect both a broader view of the landscape as a complex social phenomenon, and current political correctness. Most striking to me is the invasion of the "ordinary landscape," those landscapes not professionally designed, into courses that were once solely devoted to high, or elite, design. As with our profession and our discipline, we need to begin to consider letting go or at least focusing or prioritizing.

Our history courses are almost invariably taught by an amateur, usually a designer who also teaches studio. There are, after all, few academic historians specializing in landscape architectural matters, and even fewer savvy enough about the design process to earn their keep in the typical landscape architectural curriculum. This will probably change somewhat as the pressure for faculty with Ph.D.s continues to escalate, but the amateur historian-instructor will be part of our scene for a long while yet. This is not necessarily bad. History should inform design and vice versa. Amateur, after all, comes from the Latin root for *love*. Sensitivity, enthusiasm, and a deep conviction of the link between history and design are no small gifts. But consider that this situation would be unthinkable in the typical fine arts curriculum, the kingdom of art history specialists, and it is in fact becoming increasingly rare at architectural schools, too. Often the more fashionable the architectural school, the more specialized Ph.D. historians on its faculty. Whatever the benefits and negatives of amateurism, they should lead to some self-examination vis-à-vis our peers.

Comparison with the role of history in still other disciplines and professions is instructive. As with architecture, highly specialized doctoral-level expertise and defensible turf are extreme in many disciplines, particularly in the humanities. The common organization of literature in the university consists of watertight isolation by genre and period (and often by "culture" today), a triumph of format and chronology over meaning. In contrast, history plays no significant role in the education of engineers and bench scientists. When the history of engineering or the history of physics is taught, it's usually a seminar for honor students. Recently, some socially conscious medical schools have introduced courses in the sociology of medicine courses that often have historical aspects. In law, history is a rare, arcane specialty, set aside at the LL.D. level for probably one lawyer in hundreds. But if course catalog listings would indicate that the study of legal history is nonexistent, in some ways our own system of common law *is* the study of history, in which precedents *make* law. In the social and behavioral sciences, such as anthropology, sociology, and geography, history is commonly taught as an intellectual entry into concepts of the discipline for graduate students.

How then did history become an indispensable requirement for education in our discipline and simultaneously a subject that almost anyone can teach when faculty rotations are required? The simplest answer, of course, is continuity from Beaux-Arts education. Landscape architectural education, at least in the design-linked genealogy, began in those Beaux-Arts academies. Architecture has done more than we have to question or change the place of history in the schools, probably partly because of its larger mass and the development of architectural history as a specialty. History, not just historicism, was considered as an oppressor by the radicals of the Bauhaus. When modernism became convention in architecture schools, the forms of the European masters were quickly adopted by the more progressive architectural schools but not their political or anti-historical philosophy or polemics. We teach history because we somehow know it's valuable and because we took it when we were in school. There are probably worse reasons, but the questions remain: *what* history should we be teaching and *why?*

Commandments for Any History We Teach

Let me offer a few personal commandments relevant for *any* history we teach. I am discussing the undergraduate history that is instrumental, a functional part of education to be a professional or pedagogical landscape architect. I am not dealing here with courses preparing one to be a scholar of landscape history, nor am I speaking about liberal arts courses, important as they are.

Our History Must Make a Difference. The history we teach must be relevant to design. It must make the students reflect upon decisions that they make in the design process. The correspondence of John Evelyn *per se* is a subject for scholars but not for the studio designer, but the availability and use of plants that it reveals *is* of relevance to our design history.

It Must Not Be BAD History. This role I propose for history, isolating it from specialized work, can also isolate it from the knowledge of informed scholarship. While history does not have to deal with each current nuance of scholarly study, it must not be simply wrong or superficial, passing on lore of decades ago that is currently rejected by or questioned by scholars. Often it does. Other habits often lead us into teaching bad history also. Eclecticism might be one of our strengths; naiveté is not. In our enthusiasm to cover the breadth of landscape study, we often turn to currently fashionable books of dubious authenticity. A book becomes hot from many sources, from *Newsweek* for the least sophisticated of us (remember Alvin Tomer), and for those precious few of us, the *New York Review of Books.* We have bought into some very bad and/or irrelevant stuff. Before we accept currently popular books in anthropology, sociology, popular culture, or whatever, we should ask ourselves whether we would give Charles Moore et al.'s *The Place of Gardens* to architectural students as an example of the best in contemporary landscape design thinking.

Serious Questioning of Our History Is Different from Trendy Revisionism. Landscape architects are too genuine and too naive to pass on some of the absurdities of revisionism in the current art world, but our good intentions often do produce an overlay of trendy phrases, concerns, and disclaimers. Our canon is biased and arbitrary. Whose is not? Our job is to encourage our students to ask questions of it and to develop their ability to do so. It is not to ask those questions to ourselves and then supply the students with currently fashionable answers. We should turn to other traditions, not for models, not for vocabularies, but as stimulus for reflection on our own habits. Literary criticism, for example, is of less value to us for its content than for its sense of important intellectual cycles, cycles that we are unaware of in our own work. Terry Eagleton offered this kind of valuable insight in noting that the last 150 years of literary criticism could be divided into roughly three equal periods. The first was biographical and concentrated on the author, the second structural and concentrated on the text, and the third contextual and concentrated on the reader. Are we still mired in that first phase?

Beware of the Reification of Culture. Contemporary anthropologists point out that culture is not a real phenomenon but a mental construct that we have erected to help us analyze the world. "Culture" is neither a monolith nor a black box. Society arrives at a culture and ideology as a product of interaction and transaction amongst agents, interests, and institutions. Generalizations about culture and the relationship to landscape are not necessarily bad, if they are phrased with caution and as interpretations. Too often when we refer to culture, we are referring only to the collective taste and ideology of the power structure and patronage that sponsors art and shapes the most visible landscapes.

Whenever Possible, Study Landscapes That Can Be Experienced. The essence of landscape is change, a fact we find hard to admit. Studying real landscapes brings us face to face with their flux and their messages. It keeps us honest. Looking at landscapes that are out there, evolving and changing, also allows us to draw insights from our students. The landscape they see is often one to which we are blind, but it is just as real. Working in a real landscape gives their insights a better chance to vie with ours.

Be Much Clearer about the Relationship We Posit between the High and the Ordinary Landscapes, between the Designed and the Common. I am struck by the introduction of the vernacular (common, ordinary, pop, whatever) landscape into our history course. As someone who has made his intellectual living off this subject, I can hardly object to students studying it. But I am not clear myself what lessons for designers are to be learned from the changing "high/ordinary" relationship. I can think of many reasons for looking at the ordinary landscape in and for itself, but why within the standard History of Landscape Architecture course? The long-standing dichotomy between designed and non-designed, or between high and ordinary (two very different distinctions, note), is not only ill-defined but simplistic, misleading, and often false. The concepts behind these fuzzy distinctions could potentially lead to powerful insights about the social context of land and building. But it is not simple. Neither is it a subject for this essay, but for a future one. What for now I would ask is that teachers be clear and explicit, with themselves and with their students, about *why* certain landscapes are productive for study, be they professionally designed or not.

FUNDAMENTAL PRINCIPLES FOR OUR HISTORY

Meaning Is As Important As Form but Is Not Inherent in Form. Meaning does not come from a form. A form acquires meaning because of its place and function in a set of social or institutional values and systems (and sometimes, of course, personal history). These can be as broad as the Catholic Church of the Counter-Reformation or as narrow as a small clique of think-alike designers. Meaning is shaped and realized in a form; it withers; it is destroyed; it is often revised or reconstructed. Sometimes landscapes outlast the institutions and values that gave them meaning. Landscape, unlike literature, is not optional. Landscapes are not buildings either: they are more flexible, their interpretations wider, their access broader. Different groups, even different individuals, can read much different meanings into the same landscape. What meaning is recognizable or common to any significant group can be used and manipulated. The larger landscape and the larger society are not just context for our monuments; rather our designs are their creatures, even though either can outlive the other.

History Is Best Thought of As Lumpy Change. It is a constant flow, varying in velocity and width, replete with streams, locks, dams, waterfalls, and swamps. Change is the essence, but change is not even. (I owe this observation, but not the clumsiness of the prose, to Peirce Lewis.) The process of teaching history, or writing about it, or using it in the studio, is the process of freezing, slicing, and abstracting bits, chunks that we define. How was the Renaissance landscape experienced by those who lived in it? Was our concept of renaissance even meaningful to them? In 2050, will the Donnell Garden and Harlequin Plaza be considered part of the same historic style? Freezing time, stop framing, is a powerful tool of analysis. It is also much like the process of design itself; a building or a garden (less) is itself a freezing of time. But the stuff in between can be just as important. We constantly note, and deplore, the pace of change, but somehow fail to accept it as a landscape given. Change, indeed, is the essence of landscape, whether almost glacially slow, or so rapid as to defy understanding or control, as it is now in much of the urbanizing third world. Abstraction from time might be necessary for teaching and learning, but it needs not to be confused with reality. We might all ask ourselves whether the history we teach is more like an album of still photos, a carefully edited film, or a kind of video verité.

While History Might Be Made from Massive, Irresistible Forces of Near Universality (It Might Not Be Either), Those Forces Are Given Concrete Shape by Local Contingencies. Nowhere is this truer, or as central, as in landscape. Landscape, as Jackson has noted, is human history on the land. It is local expression of larger than local forces. Those forces, today, are often far beyond the control of the shapers of any particular landscape, but their impacts on a particular place are distinctive. The common, the ordinary, and the monotonous reflect the broadest of cultural forces. But variety is also key, not for the sake of variety in itself, but because historical change takes place in a specific landscape. We need to study the local and the distinctive as well as the universal and the dominating.

All these general remarks pertain to any history we might teach. They still do not answer the *what* and *why*. What history do we teach? How do we arrive at that decision? I have no prescriptions, but I can suggest an approach and some suggestive examples.

Three Alternative Directions

The History We Teach Must Depend upon Who We Want to Be. If our history courses are instrumental to our roles as professionals, then they should sharpen our performance in those roles. This is equally true of learning from history. The questions we pose, too often dominated by convenience, the conventions of academia, and current trends, should be relevant to what we need to know to do our job better. What is that job? We should reject the chimera of renaissance, gender-free, captain of the design team and speculate upon how history could support more focused roles for a landscape architect. Three roles worth exploring come to mind: the landscape architect as a *form giver,* as a *professional* embedded in a society, and as an *intervener,* a manager of change upon the land. These roles lead to a history of form, to a social history, and to a history of landscape change.

Take the role of *form giver* first. The appropriate history might concentrate on works now in the canon, but it would approach them differently. It might begin as a conventional studio problem, by setting up the program given to the original designer. That program is not always directly available, but piecing it together would itself be a learning experience. The next step in the process would be to link program to solution, to critique the match, and to develop alternative design solutions to the same problem and then to evaluate them. If, instead of endless series of beautifully crafted chipboard models of Villa Lante, we developed alternative solutions through computer graphics, in effect staged a design competition with feedback, what might we learn? But many other questions would be asked about the development, the construction, and the life of any project. What historic formal prototype solutions were available to the designer? What was known and what not, what drawn upon and what rejected? Was the commission and the resultant landscape typical for its time? Did it in turn become a prototype, or was it a breeding anomaly in the evolution of form? What materials went into that landscape, what crafts and what types of labor built it? How did the joinery work? What was the riser-tread ratio? (What could we learn about analyzing riser-tread ratios, à la Fletcher Steele, from the landmarks of landscape architectural history, from Villa Lante to the great pyramid at Tikal?) What were the grades of the paths at Stourhead? How was a landscape cleaned, maintained, and refurbished? What were the original plants? In the case of the Alhambra, for example, no one knows, but what were the *possibilities* for those plants and for their location and installation?

What plant palette was available to the designer? Were those plants "native," or the products of a regional or transregional commerce, or were they brought by a conquering group? How might the planting have changed over time and why? A comparative analysis of the four or five works that would be so studied would culminate that semester. This intensive study could create not only artists and craftsman but connoisseurs of the landscape art.

Another vision of the landscape architect is that of a *professional* whose work both serves and expresses the values of the society in which she works. While the product is still form, we would study social context of the design in order to discuss "meaning." The emphasis would be not on what forms have been produced but how those forms served a society and attained meaning within a given context. How can one understand why the Alhambra took on the shape and text it did without understanding the culture of Moorish Islam and the system of authority in palace, household, and harem? If the first path is a history of form, this is a social history or a history of meaning. It views the landscape not as a one-off tour de force but as an artifact created, maintained, and evolved in a cultural context. As the first history concentrated on a few built designs, so would this. But this course would also focus on a period or a genre. We would pick periods or types in which values and social organization, and hence meaning, were in flux. One example is the English garden from Stourhead to the formal-versus-natural debates at the beginning of the twentieth century. This history would take us from country estates to crystal palaces, from the refined taste of a handful of Augustan esthetes to the rise of the industrial rich and the new bourgeoisie, to imperialism and colonialism, to exploration and plant discovery, collection, and classification, to plate glass and moist heat and cast iron, and through the changes of meaning connected with such social and technological transformations.

Or, one might study Central Park as a design, as a public pastoral interlude, and as a creature of politics, taste, and social values, including those of class, moving from the original program controversies onto such "current" issues as personal safety and fear of crime, both originally concerns of Olmsted himself.

This course must take examples from cultures and societies contrasting with our own, for example, the Wang-Shi. Inquiry beyond our own tradition would highlight the various ways in which form attains meaning under varying combinations of economics, ideology, technology, and social hierarchy. It would also expose the filters with which we have always viewed such work, filters peculiar to our own society, filters that change as our society changes. Few forms of landscape exhibit this more clearly than the Chinese Scholar's garden. Any intelligent, reasonably skeptical reading on the thousands of pages of materials that Westerners have produced could show how little we *really* know about how the users of these gardens perceived them, valued them, and felt about them. The elevation of a few examples of scholars' gardens as shorthand for almost two thousand years of an art form in such a complex and highly developed society is indicative in itself. The same forms and arrangements have been dismissed by Westerners as vulgar, silly, and lacking order in one period only to be romanticized and rhapsodized over as "a beyond rational unity between person and nature" in another. And is it really likely that Chinese gardens changed hardly at all over those centuries? Fiction and literature might be just as useful as scholarly analysis in the treatment of these works, showing us not only how the perceived meanings of gardens change across cultures and centuries, but showing that any interpretation tells us as much about the culture making it as the culture "studied." This is necessary prelude for richer, deeper understanding.

A third vision of the landscape architect is expressed in the subtitle of this journal, "The Planning and Design and Management of the Land," all of which means *intervention*. The most significant mark of our contemporary landscape is rapid change. Establishing a fit between human and natural systems, then, means guiding and managing landscape change. So this landscape history would concentrate on place change. We would view history not as a series of incidents, but a process of change. Process would come first, products second. While much change is globally driven, the study of such forces is the province of social and economic and political historians. But, as mentioned earlier, no matter how strong the global forces of change, they take their particular form within specific landscapes. As they modify those landscapes, so the landscape and the local culture modify the products of that change and make them specific to a place. If we believe this, then, we could most productively con-

Staked vines in spring, Napa, California. (Photograph by Brenda J. Brown, mid-1990s)

centrate on the history of particular landscapes, from geological and vegetative origins, to aboriginal inhabitants, to the process of different groups in societies moving through and shaping cultures, on down to our post-modern, almost placeless being upon the land. Here again, cross-cultural comparisons would deepen our understanding.

Take vineyards as an example. Wherever in the world we find the classic category of a Mediterranean climate, characterized by warm moist winds in winter and dry summers, grapes have been grown and wine made. The landscape of vine and wine is as intimate, compelling, and demanding a bond as exists between human and nature, and as place specific. Not only can wine grapes be grown in only certain climates, but every micro-aspect of the site affects their quantity and quality: long and short term temperature and rainfall, steepness of slope and its attitude towards sun, the local soil, the micro-drainage, the presence or absence of moisture at night, and so on. Grower and vintner learn to work with what they cannot change and manipulate what they can. Terracing, trellising, planting, spacing, and pruning all become more than science, more than craft, and they transcend into art. And as all landscape-people relationships are to some degree indeterminate, the grapes and the growers are subject to the vagaries of nature.

What better case study than such landscapes? Think of the Napa Valley, product of our own West and its myth of California as lotus land. Mountain boundaries open to the bay at the south, sequential occupation by American Indians, ranchers, by the rich retreating from San Francisco, wine growers, wine makers and, since the early 90s, sadly and apparently triumphantly, urbanization. It's all there, from splendor of the terrain to the enormous variability of micro-econiches, from functional vernacular to post-modern monuments, from conflicts between grape growers and wine makers to the politics of development and environmental response that make strange bedfellows and new landscapes.

Charlemagne's Vineyard, Burgundy. (Photograph by Kenneth Helphand)

We could compare Napa with other wine making regions of the world, say the Medoc and Burgundy. Three different regions, committed to intensive wine making for fifty, two hundred fifty, and seven hundred fifty years, but developed under different political, economic, and transportation regimes. California emphasizes process, technology, and science, the others the fruit and the spirit of place, or *terroir*. Now all three resonate one to the other in technology, trade, and craft, all in a global culture overlaying unique places.

The way to learn about the human landscape is to study it in terms of people in a place and not monuments, styles, and periods or even "honest vernacular." A final advantage of studying the history of change in specific place is that the relationship between the "high" and the "ordinary" can be more productively explored, for the relation between them, however defined, is first of all a matter of social and physical context. If landscape is the imprint of human history, it is too important to be left to antiquarians and genealogists of the county historical society.

It is probably obvious that this third course is the one that I would choose to teach.

Conclusion

Each of you might try inventing your own such course. I know that some of you already have. The courses I suggested overlap and share common material. Maybe the distinction between them is not always clear. But the essential difference is that of a single conceptual focus on how to understand a landscape, even what a landscape is. It is that focus that gives a framework (not a theory) that fits facts together into an integrated understanding. The point is not to advocate these three particular courses but that our history courses should depend on what we want to do. If we can't decide what we want to do, or think we can do almost everything, our history courses will reflect this, to our detriment. And just as our own history courses should reflect our view of the profession, so should we evaluate the relation between history as an instrumental course aimed at future professionals, and landscape history as a liberal arts course aimed at a broad range of students. To assume that the same course can serve both audiences, as we commonly do, is a recipe for failure.

Our history courses must have rigor. They must not be "bad" history, but scholarship that can be tested and built upon. They must have a focus appropriate to what we see as our role. Such focus requires conviction, but need not entail bigotry, intolerance, or revealed dogma. (My history under Alfred Caldwell suffered from all three, but proved to me that fanatical beliefs beat no beliefs at all.) Students should be encouraged to question the tenets and assumptions of the course and to discuss them. Exposure to strong views, accompanied by a demand to think and to respond, are more likely to produce mission and direction than are permissive and passive recitals of designers, dates, gardens . . . and a bit of environmental consciousness.

Developing such courses is a daunting task. Some of the material (not much) is in our hands. Other parts can be gleaned (*very* respectfully, please) from other disciplines. Other parts we ourselves will have to investigate. Above all, we will have to integrate and focus. But in a time when publishing seems more important in the tenure process than teaching, designing, or building, in a time when such publishing is often biographical minutiae or rephrased descriptive geography, isn't this a challenge for real scholarship for a relevant mission?

SUGGESTED READINGS

On land and history . . .

Braudel, Femand. 1988. *The Identity of France*. Vol. 1, *Land and Environment*. London: Collins.

Jackson, John Brinckerhoff. 1994. "Southwest." In *A Sense of Place, A Sense of Time*, 13–67. New Haven: Yale University Press.

On larger forces and local change . . .

Lewis, Peirce. 1972. "Small Town in Pennsylvania." In *Regions of the United States*, ed. John Fraser Hart, 323–51. New York: Harper and Row.

Raup, Hugh. 1966. "The View from John Sanderson's Farm." *Forest History* 10 (1): 2–11.

On wine regions . . .

Conaway, James. 1990. *Napa: The Story of an American Eden*. Boston: Houghton Mifflin.

Loftus, Simon. 1993. *Puligny-Montrachet: Journal of a Village in Burgundy*. New York: Knopf.

And on vines and wines, on mind and land . . .

Kramer, Matt. 1992. "The Machine in the Mind." In *Making Sense of California Wine*, 31–58. New York: William Morrow.

———. 1990. "The Notion of *Terroir*." In *Making Sense of Burgundy*, 39–48. New York: William Morrow.

Seeing the Unseen and Listening to the Local

RACHEL LEIBOWITZ

Bob Riley was not an active preservationist. He thought about landscape in profound ways and was an expansive reader of divergent texts, both fiction and nonfiction. Bob had a clear appreciation for history, and I think he had sentimental bones. I know that he was truly awed by the capacity of landscapes to foster multisensory experiences and to jolt memories. However, he was not a proponent of historic preservation. During conversations in which I would mention a current proposal to alter or even demolish a particular place, he sometimes would get excited and say "Ah!" before sharing his own fond recollections of it, along with his concerns for the proposed changes. Other times, though, he would seem to say with a gesture or a shrug, "*C'est la vie*," the show must go on, you know that everything is in flux. To Bob, the practice of preserving not only history but also place sometimes seemed a bit too precious.

Surprisingly, perhaps, in my work as a practitioner and educator in the field of historic preservation, I sometimes find myself agreeing with that perspective. I often have my doubts that we need yet another house museum that tells the story of yet another founding family whose name is well known within a five- to fifty-mile radius, no matter how exemplary its framing, its brickwork, its gingerbread. So much of preservation seems to emphasize the object, to focus upon form, upon rarity—and to dismiss the commonplace, the ubiquitous, the too-familiar even in the histories of our everyday lives.[1] The sharp focus of too many practitioners of design—architects and landscape architects alike—and of preservationists is trained upon what we may see (surface and style) and not enough upon what we may know (histories and contexts). This is how many of us were trained, and it is how we train future practitioners. But is that repetition out of convenience? Or perhaps of reassurance, as if to say, "We all know how you want places to look and how you want events to be remembered, and we have a recipe for you to follow"?

To these questions, I would answer, simply, "yes" and "yes." Not a nuanced response, but my reply is based upon some of my own experiences in the field. I have seen the bureaucratic, box-checking approach taken to its limits far more often than I have seen creative, open-minded, and open-hearted approaches to stewarding our historic built environment. Political alliances and economic strategies are difficult forces to overcome; buddies and budgets wield more power, typically, than folks' fond recollections or even the most deeply revered local histories and unusual, treasured landmarks. And then there is the unstoppable force of consumer culture, perpetually seeking the thrill of the shiny and new at the direct expense of the existing, as though it were the existing that has prevented us from achieving the shiny all along. There is a common perspective among the general public that "hysterical preventionists" want to stop all change, opposing all bulldozers along with any politicians, developers, and homeowners who seek to live in the present or look to the future. Yet in fact, historic preservation links the past to the future through the present.

"What History Should We Teach and Why?" in Preservation Practice

It was only as I sat in Bob's classroom in 1996 as a new architecture student that I first started thinking about the design of the built environment as a whole, as a complex system. Until then, I had explored buildings but not what surrounds them. The architecture school faculty, designers and historians alike, framed celebrated buildings as high-style objects in space; even historic gardens and urban plazas were presented as formal exercises and mere backdrops for buildings.

Bob's class was different. He encouraged us to go outside, to look at spaces between buildings, to pause in front of every house and look closely at its front yard, its relationship to other houses and the street. How did the penchant for an enormous, decorative boulder near the mailbox get started, he asked? What is happening in the narrow alleys downtown? He challenged us to find the places where the city blurred into agricultural fields. Is there a border, an edge? Can you see it? Can you hear it? From Bob, I learned that there was so much in the built environment to discover and wonder at, and that all of it—even the most humble—should be investigated and studied with rigor.

In his 1995 essay "What History Should We Teach and Why?" Bob challenges the typical instruction of landscape history. Directing his remarks towards educators and instructor/practitioners in departments of landscape architecture, Bob "offers some cautionary commandments" and "suggests basic principles" for the study and teaching of history, among which two stand out: "Our history must make a difference," and "Our job is to encourage our students to ask questions . . . and to develop their ability to do so."[2] Elsewhere in the essay, Bob suggests creating a landscape history course for students who would choose to guide and manage landscape change at a local level. Although it may not be precisely what he meant, that is the work of a historic preservation practitioner. The role of a historic preservationist is to engage with communities, to learn about landscapes from the people who care about them, and—if asked—to help guide their decisions to manage their changing landscapes.

Architectural Privilege: How Something Becomes "a Thing"

In the state historic preservation offices in which I have worked, most staff (excluding archaeologists) emphasized the visible—what could be apprehended and understood by sight onsite. This is due, in large part, to what I think of as "architectural privilege" in historic preservation practice. Historic preservation office staff and consultants submitting their work for staff review have also been trained to prioritize building form and style. When staff determines whether a building is historically significant and eligible for listing in the National Register of Historic Places or in an analogous state register, their focus quite often is on the building's form, material, and style, whether "high" or "low." The surrounding site context or "landscape" often does not even enter the analysis. The approach to determining appropriate scopes of work for historic buildings is grounded in first examining existing conditions of the building, not its site or context. Therefore, it is unusual to find staff in most state historic preservation offices with a background or professional degree in landscape architecture.

Throughout the country, these offices are also responsible to review federally or state-funded or permitted projects. Sometimes those projects involve high-style, iconic buildings like a courthouse, but other times they affect underappreciated resources of infrastructure, whether bridges and viaducts, water towers and filtration plants, or modestly designed parks, forest preserves, and other "open" spaces that seem to lack any visible resources whatsoever. For these undertakings, as in the building rehabilitation projects, emphasis would be placed upon form and style, relying upon the consultant's or staff reviewer's eye to tell if the resource in question was, indeed, "a thing" worthy of preservation.

The guiding light defining the "thingness" of a resource—its relative rarity or its ubiquity—is, more often than not, the individual's own education and training in architecture or architectural history, or readily available and easily attainable secondary-source material: for example, a popular book about house styles or the evolution of the skyscraper. If the building, structure, object, or landscape had been written about already, or a similar example in another town or state had been listed in the National Register of Historic Places, then its historic significance had been recognized and validated. However, primary sources, such as letters from a property owner or a designer, or contemporary newspaper articles, photographs, or paintings of a site, typically are not carried around in the automobiles of those conducting cultural resource surveys and, therefore, are not commonly utilized during this "windshield" identification phase and subsequent consultation process. Look at it, recognize it from a class once taken or a book once read, check the appropriate box on the survey form, and move on to the next structure; if it's not readily apparent as a thing with a particular form or style, or if it's not of a particularly obvious stature or well-established status, then it's not likely to be a thing. Sight and site (and cite), unseen.

Questions, Form, and Meaning

This is where lessons learned from Bob came into play for me. I had internalized his urging (admonition?) that even the humblest of resources is worth deep consideration, with rigor, and that mere descriptions and appreciations of form and style could take us only so far. "Meaning is as important as form," Bob wrote in 1995, "but is not inherent in form. Meaning does not come from a form. A form acquires meaning because of its place and function in a set of social or institutional values and systems."[3] Local histories can reveal facets of significance hidden in plain sight that, if shared, can enrich and enliven our understanding of the built environment and, by extension, the human experience.

So often I would read a consultant's historic resource analysis report that inventoried what should be demolished for a new highway project, or a draft of a National Register nomination, and I'd have so many questions that needed answers before I could decide for myself if the resource was a thing—an eligible historic cultural resource. More often than not, I would err on the side of the resource and consider it as a culturally and historically significant place until I could be dissuaded of my assumptions—I needed proof that "nothing happened" here. Predictably, this was to the great annoyance of colleagues or applicants who claimed to not have the time or the capacity to research further the place under consideration, especially if its condition was poor or if its aesthetics were unimpressive. But I always encouraged a deeper investigation into *why* what we were looking at was *not* a thing, or why it was being considered and nominated as a thing.

Sometimes a cultural resource doesn't make much of a first impression, like the earthen levees along the Rio Grande. When I first determined that a southwest Texas levee system was eligible for listing in the National Register, my colleagues and staff at both the International Boundaries and Water Commission and the Army Corps of Engineers were aghast. How could piles of dirt be historic, let alone significant? Recalling Bob's push always to question what we see, I responded to them with more questions. How is this *not* culturally or historically significant? How are these earthworks, engineered by the federal government, constructed by local laborers under the New Deal, and continuing to serve their original purpose of flood-control, *not* historic? What about this system along a boundary recognized by two nations is unimportant? After months of discussions and research, my colleagues and the federal agencies acknowledged that, indeed, we should treat these levees as historic structures, and not merely as piles of dirt to be replaced by concrete, and a design solution was sought to increase their height yet maintain their historic slopes and profiles. (Deciding whether or not the engineering of a river and its banks is ecologically sensitive and responsible was not our task at hand.) The meaning of this levee system, its historic significance, was not inherent in its form, but the form had acquired meaning because of its place and function over time.

Conversely, sometimes form and style can provide visual clues that suggest an important history, forgotten or repressed. When a local arts organization submitted its draft of a National Register nomination for an Art Moderne–styled movie theater in Edna, Texas, I asked them why the building's entrance doors under the circular marquee were asymmetrical. In its overall form and in its style, the building itself appeared asymmetrical yet balanced, as the Art Moderne so often did, but I wanted to know if that style was hiding something—

naturalizing something and rendering it invisible simply by incorporating it into its stylish appearance. The nomination's authors explained that the single pair of doors to the left of the ticket booth was the segregated entrance that had funneled Black patrons straight upstairs to the balcony, without access to the lobby, its glassy concession stand, and its bathrooms, while the two pairs of doors to the right were for the white patrons. Heading downtown on a Saturday night, friends could approach the neon glow of the theater together, but once they had stepped upon the radiating lines of terrazzo sidewalk under the marquee, they were separated. The authors added this information to their nomination, and the building was listed in the National Register of Historic Places not only for its exceptional architecture at the local level, but also for its social and cultural significance to the community as a place of entertainment with very different moviegoer experiences during the Jim Crow era. In late December 2019, eight years after its historic designation, the Edna Theater in all its neon glory was featured in a colorful photo in the *New York Times*, and I was reminded of those efforts to tell a deeper story of this place—which began merely by questioning the building's asymmetrical entrance, its form and its style.

An excellent example of social history meeting resistance from architectural privilege, meaning versus form—and a fascinating example of the struggle of substance over style that I think Bob would enjoy—can be found on the northeast corner of North Eighth Street and Lawndale Avenue in Pekin, Illinois. There stands a mighty but modest monument to modernity, movement, and money: the first automobile-oriented shopping center or "strip mall" in the western region of the state. Designed and built beginning in 1950 by Orfeo Gianessi, an Italian immigrant to Illinois, this building predates other commercial strips in the larger city of Peoria, eleven miles to the north.[4] This one-story row of thirteen storefronts and its companion strip of shops completed in 1958, also built by Gianessi with his architect son-in-law, Silviano Lippi, used quarry-faced Indiana limestone and a thin metal canopy with bullet-shaped spotlights to highlight the enormous display windows that allowed tenants, including the family's own shoe store, to present their wares to passing motorists. Though today it looks rather humble, at the time of its opening in the 1950s, it was unusual and attention-grabbing, more modern and exciting than anything built to date in Pekin and the Peoria metropolitan area.

When the owner—the grandson of Gianessi—wanted to nominate the North Eighth Street Plaza for listing in the National Register of Historic Places, he selected a consultant and shared important information, including original sketches and architectural plans and his family's scrapbooks filled with newspaper articles and advertisements. The cultural and social significance of the place—certainly at the local level, and perhaps at the state level—was undeniable. But some readers of the draft nomination became concerned with what they could see, which no longer seemed fashionable and looked a little worse for wear.

The state board of review felt that the North Eighth Street Plaza had merit for its social history, its groundbreaking role in local commerce related to the automobile and state route connecting Pekin to Peoria, but they did not see merit, let alone beauty, in the shopping center's straightforward, modern design. They did not thoroughly consider the building's Modernist juxtaposition of texture and glossiness in its limestone and display windows, its direct relationship to the road, its automobile-oriented landscape. They were fixated on its lack of "ornament" signaling a lack of significance rather than a connection to postwar architectural Modernism, and on the aging of its materials, which in any other style would be considered "patina." In a sense, they were blinded by what they could see—and by architectural privilege. They could approve of forwarding the nomination for its cultural significance, its social history, though invisible to the eye; but they could not agree upon nominating the strip mall for its design.

Where was the intellectual rigor in their arguments? The North Eighth Street Plaza reads precisely as it had upon its midcentury grand opening. The building had not been changed, but our perceptions of it had, to the extent that they clouded our ability to read the resource. During the meeting, the board members didn't like what they saw. They found the building to be unpretty, if not ugly, but they liked its story—what they could *not* see. It was only after the owner stood at the podium to explain his grandfather's design intentions and attempt to connect the fascinating story of the man and his vision, built in limestone, concrete, glass, and asphalt, that the state board of review understood that they were looking at something—a thing—in addition to learning about a local commercial history and a postwar entrepreneur. Instead of tabling the nomination as they were poised to do, the board asked the builder's grandson to help the consultant flesh out a more thorough articulation in support of the Plaza's architectural significance, taking a more rigorous approach to the explanation of its design origins and its subsequent influence on the built environment of Pekin. After months of additional work to connect substance to style more deeply, the meaning to the form, the important-but-invisible history to the visible place, the Plaza was listed in the

National Register of Historic Places, not only for its historic commercial significance but also for its groundbreaking design significance at the local level.

"Our History Must Make a Difference": Preservation of the Visible and the Invisible

Beginning in 1992, the US Department of Transportation designated the route between Chicago and St. Louis as a developmental high-speed rail corridor. As a federal undertaking, this would require coordination with the state historic preservation offices of Illinois and Missouri, state transportation agencies, and local authorities to manage its effects. This potential project became a reality in 2010, breaking ground with a federal grant from the American Recovery and Reinvestment Act. In the state capital of Springfield, Illinois, the project required widening the right-of-way, which would destroy the archaeological remains of a Black neighborhood devastated by violence in August 1908—the first race riot north of the Mason-Dixon line since Reconstruction.

This horrific event had resulted in the deaths of seven people and the loss of twenty businesses and nearly forty homes owned by African Americans. Very quickly after the riot, the remains of this decimated neighborhood were removed or buried with fill—resulting in a landscape of violence rendered invisible for more than one hundred years. Race riots were all too common at the turn of the twentieth century, but this riot in Springfield captured the nation's attention because it occurred in Abraham Lincoln's hometown, just blocks from his former home and even closer to the state capitol building where he had given his famous "house divided" speech. Newspapers across the country posed the question: if this can happen in the home of the Great Emancipator, then has this nation made any progress in race relations? In direct response to these questions and this event, social reformers, including W. E. B. Du Bois and Ida B. Wells, worked to establish the National Association for the Advancement of Colored People (NAACP), which was founded on February 12, 1909, the centenary of President Lincoln's birth, and just six months after the horrific event in Springfield.

Although the direct evidence of this hate crime had been erased from view, its horrors were not forgotten by the Black community and its allies. Community members knew the stories of their elders who had witnessed the event, and they kept this history alive. Its effects remained very present in the segregation of the city, physically defined by this railroad

Excavations of the former neighborhood decimated in the August 1908 race riot in Springfield, Illinois. Covered with fill for more than one hundred years, the site includes several house foundations, sidewalks, and yards located within a proposed high-speed rail corridor. (Photograph by Rachel Leibowitz, June 2019)

corridor that would now be widened. In 2014, as Springfield sought to move forward with the multi-million-dollar rail plan, it became clear to our office that the project would destroy any archaeological evidence of the nationally significant 1908 event and such loss must be avoided, if possible; if not, the public should be able to voice its ideas for what might mitigate this loss.

When an archaeological survey confirmed that the remains of these houses were still in place, the project was temporarily halted. The Federal Railroad Administration's preservation officer, an archaeologist by training, visited the race riot site and determined that it was eligible for listing in the National Register at the national level of significance. The high-speed rail project was paused to comply with federal preservation laws that protected the archaeological resources while local stakeholders discussed their desires for the future of both the site and the transportation project.

One of the markers—an interpretive monolith with recognizable logo standing in a triangular sheet-metal planter—along the Springfield Race Riot walk dedicated in 2018, the 110th anniversary of this horrific event and the state's bicentennial year. This marker is the most poorly maintained and least likely to be visited by tourists; standing near the corner of Madison and Eleventh Streets, it locates the site of the lynching of Scott Burton, a fifty-six-year-old barber, on August 15, 1908. (Photograph by Rachel Leibowitz, June 2019)

The mayor and several state legislators were not happy, to put it mildly. Engineers would now have to go back to the boards to see if the archaeological sites could be avoided. Months of discussions turned into nearly two years of delays, with local officials and state representatives growing angrier and more frustrated. The federal funding had a deadline that might not be met; local leaders had little to no interest in telling a very difficult, traumatic history. But to the credit of all local stakeholders, engineering consultants, and government agencies involved, the rail corridor was moved by several feet, allowing five of seven house foundations to be saved in place.

After these painful discussions, the City of Springfield recognized that this important history was, indeed, something to acknowledge, learn from, and honor. In 2018, in time for the State of Illinois's bicentennial, the City quickly installed interpretive markers along the route taken by the rioters, so that the public could follow its destructive path and learn about this tragedy. This is an important step in the right direction, although the walk that tells this history could be more fully and powerfully presented with a more engaging landscape design. Perhaps most surprising and encouraging of all, the state legislators who had opposed pausing the railroad project to allow community engagement and consideration of alternatives to save the archaeology in place now strongly support its preservation and interpretation. They have asked the National Park Service to study the feasibility of designating the location of the 1908 Springfield Race Riot site as a national historic site; the Park Service has found that it is worth further investigation and has begun a higher-level inquiry.

Histories of "People in a Place"

In all of these examples, community members played a prominent role in these tremendous preservation challenges, and they are owed the greatest thanks and appreciation. None of these projects could have been accomplished without their deep knowledge of place and local history, their conviction and determination to see the memory of the past advance into the future. Yet credit and deep thanks also must be given to Bob Riley. He knew of the importance of learning about places from the local people who know them best—those responsible for maintaining earthen levees, or multiple generations of moviegoers, or the descendants of immigrants who dreamed of new opportunities, or people who cared deeply for a place hidden from view, a place that appeared empty and available to others who could not see it, who did not know it as they did. Their work made these historic landscapes visible to all of us.

Bob urged us to study landscapes "in terms of people in a place"[5] and told us not to focus merely on the great monuments or icons—certainly not without truly trying to understand their creators and how their significance or meanings have changed over time. As Bob so clearly wrote, landscapes are not buildings; they are more slippery and challenging to grapple with when considering their history and relevance. Landscapes change over time: their form changes, their contents wither and die, they are destroyed, overwritten, reinscribed. Their meanings and significance change as their uses and users change, and often these meanings are highly contested among people: "History is best thought of as irregular change, as a constant flow, varying in velocity and width. . . . Change is the essence, but change is not even. . . . Change, indeed, is at the core of landscape, whether almost glacially slow, or so rapid as to defy understanding or control."[6] To gain a deeper understanding and, perhaps, better control of change, we must ask questions of what we are seeing (and not seeing), seek answers with rigor, and always attempt to understand the meanings behind the forms. Our history must make a difference.

Bob's charge to landscape historians and educators is to make history relevant and useful, depending upon the careers that our students would like to have. Whether or not our students plan to work within the broad field of historic preservation, I believe that preservation concerns and challenges should be a part of *every* design studio that we teach. All design projects—whether in our classrooms or in the real world—must consider what existed, what exists, what can and should be changed, and what should be maintained. There is no *tabula rasa*. This is what we do, this is what we are *obliged* to do, and this is what our students will do as form givers, as professionals embedded in society, and as managers of change. "Preservation" is neither rarefied nor elitist. It is not prevention. It is engagement. It is embrace. It is for everyone who cares about where they live and its value. We can debate what history we should teach and why, but we cannot argue that designers must care about history, especially if they want to be relevant in it and make positive change in our world.

NOTES

1. Unless, of course, the sheer volume of a ubiquitous type is itself rare or demonstrates an exemplary quality, as in the breathtakingly massive quantity of humble, repetitive bungalows (1,265, to be exact) that constitute the Central Berwyn Bungalow Historic District (of suburban Chicago, Illinois), listed in the National Register of Historic Places (NRHP #15000521) in September 2015.

2. Robert B. Riley, "What History Should We Teach and Why?" *Landscape Journal* 14, no. 2 (Fall 1995): 220–25; quotation from revised version in *The Camaro in the Pasture: Speculations on the Cultural Landscape of America* (Charlottesville: University of Virginia Press, 2015), 34–35.

3. Riley, *The Camaro in the Pasture,* 36.

4. Peoria did not begin to construct its first shopping center, the Sheridan Village, until after its developer and his architect, Marvin Goodman and Dean Duboff, visited the North Eighth Street Plaza in Pekin. See Kenyon and Associates, "North Eighth Street Plaza," 2010, typescript, Gianessi-Lippi Collection; cited in the North Eighth Street Plaza National Register of Historic Places nomination; listed in May 2015 (NRHP #15000226).

5. Riley, *The Camaro in the Pasture,* 41.

6. Ibid., 36. On the reinscription of landscapes over time, and the trendy overuse of the word "palimpsest" to describe such, Bob wrote: "Most palimpsests turn out to be illegible. Finding a palimpsest does not mean that you are reading it correctly, or maybe at all. . . . And if history is a palimpsest, then who gets to erase our designs, and how soon?" Robert B. Riley, "About Palimpsests," in *The Camaro in the Pasture,* 120.

Robert B. Riley, Illinois.
(Photograph by Paul Lettieri, 1975)

III. Rural Landscapes

1985

Square to the Road, Hogs to the East

ROBERT B. RILEY

Most of us have read the tales. The quiet, the bright flowers, the tall grass moving to the wind like the sea and high enough, it was said, to hide a man on horseback. Even today, with that vast mesic grassland long plowed up, the traveler driving west on Interstate 74 senses the land open up just past the Indiana line. Here the stretches of tall grass become not an incident within the woodland but the landscape itself, a strange landscape, a landscape so different that a word for it had to be borrowed from the French. A word that George Stewart wrote that ". . . has always borne a touch of strangeness and poetry. In English, *meadow is to prairie as* a placid cow [is] to a shaggy buffalo-bull."

The strange beauty of the great grasslands has been reduced to commonplace by the rhapsodies of writers and tourist developers who never saw it. The story of its development over the years from 1830 to 1880 has been thoroughly documented. The settlers left the stream corridors, moving first to the forest prairie edge, then cautiously onto the drier prairie, while keeping a small woodland parcel for fuel and fencing, and finally out on to the wet prairie, where one could take a flat-bottomed boat for miles in the spring. The technological keys are familiar: the self-scouring plow to cope with the sticky spring soil; the railroad to open markets for the change from a subsistence to a cash economy; the osage orange hedge and then barbed wire to end the fencing controversy that dominated agricultural discussions and investment for much of the nineteenth century; and intensive mechanization to take advantage of an opportunistic cheap land, expensive labor, ample markets and maybe

Drawing by Susan Wydick, commissioned by Robert Riley, 1985. (Courtesy of Susan Wydick, from original publication)

a native inclination for tinkering. Less documented are the skills and technology of the immigrant Frisians who drained the wet prairie and the spontaneous, synergistic development of that Midwestern system of grading, marketing, and storing grain—symbolized by the prairie elevator—that was not to be introduced into the other great grain growing areas of the world for a half-century. All evolved within that uniquely American framework, the mile square grid.

All the elements were in place by 1880. The following five decades witnessed the perfection of the system. The ancient northern European animal and small grain agricultural system reached its culmination in a mechanized and highly capitalized crop rotation of corn, oats, and hay. with the raising of poultry, hogs, beef and dairy cattle, apple orchards, and miniature vineyards, and a structural assemblage of farmhouse, shed, outhouse, hog and chicken houses, ice and pump and cob houses, barn, shed, corn crib, and windmill. The image lingers still, an apotheosis of American society and settlement system, a subject of endless personal and commercial nostalgia. Each fall, the *Chicago Tribune* still runs the cartoon "Injun Summer," memorializing the quintessential Midwest landscape.

Cartoons notwithstanding, much of that landscape is gone, even in "Tribuneland" and certainly in that part of the "cornbelt," east central Illinois, where I live. We all know that, too. If we cherish the image, we sense that reality is different and growing more different. We talk of agribusiness, corporate farming, runoff, and monoculture.

The underlying technological forces that have changed cornbelt farming can be described simplistically as an electro-petro-chemical revolution. Electrification was slow in coming to rural America but finally came: dramatically and suddenly, almost entirely a creation of Franklin Roosevelt's second term. By World War II it was essentially complete. The 1930s and 1940s also witnessed the disappearance of horse and mule. By 1950 electricity and the internal combustion engine had triumphed, chemical fertilizers could be substituted for manure, and chemical herbicides and sophisticated corn genetics continued a productivity push that began in World War II. On through 1980 farm equipment became constantly more sophisticated, and more expensive, and the dense net of township roads was well graded and pawed and supplemented by the interstate highway system.

Electrification, gas- or diesel-powered tractors, and the disappearance of horse and windmill might read like ancient history; if the image of the old farm is strong, it also seems that of entirely another time and society, as sentiment increases distance. But the older, working farm couple of today began farming in that seemingly so far away time, picking corn by hand and throwing the ears in a wagon. If the last five decades have seen dramatic change, it is well to recall that people have lived through it and adapted to it.

The result of all these changes is, in simplest terms, a capital-intensive, two-crop, cash-grain system. If one took the 1929 census of agricultural production for an eastern Illinois county and replaced every product but corn with soybeans, the result would be close to the 1979 census. In those fifty years acreage in corn has oscillated about a steady level, wheat

Drawing by Susan Wydick, commissioned by Robert Riley, 1985. (Courtesy of Susan Wydick, from original publication)

Drawing by Susan Wydick, commissioned by Robert Riley, 1985. (Courtesy of Susan Wydick, from original publication)

acreage has sunk to a vestigial amount, oats have disappeared along with the horses they fed, and beans compete with corn. Sheep, hogs, horses, and cattle have all sunk to a number less than ten percent of that reported in 1929; orchard trees and vines are no longer even enumerated. John Fraser Hart has labeled the result the "CBM agrisystem": corn, beans, and Miami.

Such a description conjures up images of factory farms, bare fields and silting stream, and a stark repellent landscape, but the easy images do not always survive a careful look, and some myths are simply that. This is not corporate farming, not a creature of large companies. The investment return, well under five percent, is too small and too unpredictable to interest big business. Farming here is expensive, certainly. Two decades ago J. B. Jackson observed "that the young American without means can sooner hope to be president of a bank than the owner of a working farm," and the writers of the column "Profit Planners" in *Prairie Farmer* recently advised a would be farmer with experience and $130,000 that he hadn't enough capital to even consider getting into the business. The operative catch phrase is "cash flow," but almost all of this farmland is still owned by or among families or individuals, however large their assets might be.

Everyone "knows" that farm size is increasing, and the agricultural census confirms that lore. But for census purposes, "farm" means all the agricultural land under a single "operator." In the eastern cornbelt, a farm of large size is likely to consist of noncontiguous pieces of land separated not only by other people's land but by paved constraints of the ubiquitous mile-square township roads. Fewer farms mean somewhat fewer buildings, but beyond that the direct effect of farm size on the look of the land is not so direct or obvious. While farm "tenancy" has increased over the last fifty years, the effect of that change on the look or the economy of the land is not obvious either, although the increases in cash leasing are thought to bode badly for conservation practices. Most farmers own some land and rent additional land. The idea of tenancy conjures up images of barefoot poverty and mean living, but the *Prairie Farmer* reports that progressive tenants, seeking more land to maximize equipment use and minimize cash flow, are printing brochures that emphasize their computer-based accounting systems and, in one case, advertising in the *Wall Street Journal*. Nor are Arabs or other foreigners gobbling up Illinois farmland. Less than one-half of one percent of the state's farmland is registered in foreign ownership, and much of that ownership defined as foreign is simply the land of oil and coal companies with a majority of foreign stockholders.

But if not all lore is true, still the look of the land has changed in fifty years. Probably the most radical visual transformation, and the least remarked upon, is the night landscape: mercury vapor lights, first distributed free by utility companies in the 1960s and called security lights, cast pool of harsh, garish light at every farmstead.

What woodlots still exist away from the stream corridors are sparsely treed and clear of understory; they have invariably been grazed and lumbered over the years, hogged and logged as it is called. What scraps of good prairie remain often can be found along railroad lines or in cemeteries. Good stands of prairie are most likely to remain where a road parallels the tracks, leaving a width of undisturbed soil wide enough for some ecological stability but too narrow to be worth farming. Such strips are disappearing as railroads turn from burning to spraying for right-of-way maintenance, and economic pressure makes corn or bean raising more likely. Cemetery prairie is most likely on fenced but little visited areas, subject to enough mowing to kill off woody invaders but not enough to drive out the more fragile native species. A mowing once a year for Memorial Day (the "Decoration Day" of my youth with memories of peonies and rusting G.A.R. stars) is often just enough.

The landscape is more open. The orchards and vineyards are gone. What few hedgerows remain have gone to canopy, leaving the eye level view nearly clear. The fences have almost disappeared; neither corn nor beans wander into the road to be hit by cars, and the few animals left learn to avoid that single strand of almost invisible electrical wire on thin metal posts. Two or three fence posts remain, but only for sight lines to identify property boundaries when working the fields and maneuvering equipment.

There are fewer buildings. Consolidation of ownership usually means removal of farmsteads, consolidation into one "home place." At the remaining one farmstead, an electrified cash grain farm needs no ice house, pump house, cob house, cattle barn, or chicken or hog house. There are still over one-quarter of a million chickens in my county, about the same number as fifty years ago. In 1929 they were distributed among more than 2,600 farmers, in 1979 among less than 70, a phenomenon that serious students of cornbelt settlement systems call "fowl urbanization." Some building types have stayed but evolved. The new farmstead will have a shed, but it is likely to be lower, longer, and wider, with a roof of shallower pitch, a prefabricated industrial building of corrugated roof and siding over pole and truss framing. Where a corn crib remains, the only likely change is the replacement of wooden siding by perforated metal, an evolution of material within a stable form. But the greatest change in the working architecture of the cornbelt countryside derives from a single technological innovation: the shelling head on the combine. The word "combine" itself comes from the first turn-of-the-century field machines that both reaped wheat and threshed it but now applies to comparable operations on any crop. Until ten to twenty years ago, long after hybrid corn with its uniform height had allowed the development of a mechanical picker, corn still left the field on the cob, to remain there until fed to animals on the farm or to be mechanically shelled at the farmstead or elevator. Today in the most productive parts of the cornbelt almost all corn leaves the field shelled.

The traditional cornbelt corn crib was a clear, logical, evolutionary solution for storing ear corn and a story worth tracing in itself. Now, with shelled corn or beans or wheat, the farmer must store not a stack of loosely piled, large cobs but a dense, bulky mass of small grain particles, a mass that often needs to be dried and always needs to have air flowing through it to avoid rot. The successor to the corn crib is a round metal bin, with a fan to force air through it and often a tank of propane at hand for warm-air drying. On the largest farms there will be rows or clusters of these bins, loaded from a high, industrial looking tower, the "leg," braced with wire rigging. Such farms, and many of the newest and largest elevators, transcend the shed roof, ad hoc clustered forms of old farm and elevator to take on, with their massive tanks, complex piping, and spidery rigging, the high-tech look of an oil refinery.

But the old cribs often stay. The slatted bins can be lined to hold shelled com; the loft over the central aisle can serve as bulk storage, the aisle itself can shelter equipment. No driveway is needed because only field-tired vehicles approach them. Maybe it is not worth the trouble digging out the concrete foundations just to gain three more bushels of corn a year. Whatever the reasons, the cribs often stand alone in a quarter section of land, their stark shapes and weathered tones emphasizing the open, bare flatness of the fields and the sense of time and change on the land.

The example of the corn shelling head eliminating a vocabulary of architectural functions and forms, from crib to cob burner, and introducing an entirely new geometry, shows how little we understand about why the look of the landscape changes as it does. Technological forces affect it in complex and not obvious ways. The rural landscape for example, abounds with examples of the "cornbelt cube," a house of square plan and two stories. topped with a pyramidal roof and a central chimney, built in the early years of the century. We can guess that a major reason for its popularity was the introduction of convection flow warm air, central heating, but no one knows. No one has ever studied the American house as an expression of evolving heating technology, from fireplace to stove to convection flow warm air to forced warm air, to finally, electrical and solar heating.

Nor has anyone studied the evolution of farmstead landscapes. For decades the extension literature has offered advice to the farmer. Some of it, such as Wilhelm Miller's evocation of the prairie spirit, has been sensitive and some, like the recurring recommendation of tree massing to hide unpleasant views and frame pleasant ones, simply sensible. Too often it has been accompanied by condescending admonitions that the flow of farm youth to city jobs is due to the lack of rural beauty and would be stopped by tasteful design. A scholarly comparison of the actuality of farm landscaping with the advice of designers and improvers, however, might show that farm families have evolved their own images of "farmstead" independent of professional fashions. Certainly, the current actuality, a predilection for huge velvety carpets of grass, putting-green perfect and unbroken by trees or shrubs, would seem to owe more to the popularity of the self-powered riding mower than to the tastemakers' advice.

We know little enough about the impact of the professional literature on the countryside, but we know even less about the impact of the popular press and advertising. If fascination with the power mower and the lack of row crop work over much of the summer contribute to the bare look of the farmstead lawn, so does the advice of cornbelt newspapers and the *Prairie Farmer* to eliminate shrubs that offer hiding places to thieves intent on rustling combines. Farm-oriented advertising is big business, as intent on image as on product. Industrial type farm buildings are given names like "Ironwood," a nice blend of practicality and romance. The chemical companies have two schools of brand naming: the cleanroom, smock-coated, wonder-drug approach of Teflan, Basagran, or Cythion, and the sod-buster, tough-guy school of Bronco, Lasso, Roundup, and Fussilade, in a not surprising demonstration of the importance of both high technology and tradition in cornbelt life. This curious blend of progress and tradition can be seen in the caps distributed by the elevator. They bear the legend "equal opportunity fertilizer," but the woman's cap has a powder blue bunny tail tassel on top.

Technology and tradition have produced a different landscape, a landscape that repays care taken in looking at it for what is, not what it was. The visual changes in the cornbelt landscape can better be interpreted as continuity than as disruption. We are lucky, for this is not the case everywhere. In England, where the new farmsteads look more and more like those of America, the change is legitimately lamented. The traditional English farmstead adapted to and utilized topographic changes, formed a tight compound with buildings enclosing spaces between them, and was made up of heavy and monolithic structures. The new farmsteads sit on land graded flat with the buildings linearly arranged and made of light striated sheeting over a skeleton. This is an equally accurate description of the new cornbelt farmstead, but it also applies to the older farms here. Cornbelt farm buildings never formed a compound or utilized exterior walls and courtyards, but always appeared as independent masses. Their siting obeyed only two laws: hogs to the east and square to the road. The hogs are gone, but the buildings, new or old, are invariably aligned on the north-south east-west axes of the grid. Unsubstantial sheets of corrugated metal and prefabricated trusses are a logical extension of that American invention the balloon frame, and of the American philosophy of building lightly and quickly. The new farmstead might look simpler because it contains fewer buildings, but its visual and functional rules are the time-honored ones. Its look, stark clean, sometimes hinting of the industrial, can be seen not as a disruption of a tradition but as its clearest expression, a near Miesian perfection through reduction to essentials.

The larger landscape, too, can be understood as an ultimate expression of traditional values and beliefs. It is open, bare, clean, and organized to the point of starkness. It is an expression not just of efficiency and profit, or of cash flow, but of a belief in productivity, care, and neighborly respect. To ride a combine at harvest time, high above the corn tops, numbed by noise and vibration, is to feel not just the power in the machine but an elation in using it to transform the land. The remnant hedgerow, the old fence or scraggly tree is an affront, not just because it means a bushel less but because it resists the human ordering

Drawing by Susan Wydick, commissioned by Robert Riley, 1985. (Courtesy of Susan Wydick, from original publication)

made possible by the machine. A bare, black, fall-plowed field, however erosion prone, however necessary for planting hundreds of acres in uncertain spring mud and weather, is also seen as a sign of care expressed in neatness and order. To feel that care is to understand the affront produced by a single cornstalk in the middle of the bean field. It is to understand the state signs explaining that the shaggy midsummer ditch is actually a "roadside for wildlife," not a result of shoddy upkeep by the adjacent farmer. And to know what neighbor still means, in a time when even one-man farms average close to 500 acres, is to realize that the open, unadorned farm lawn, clean and kept like the fields around, is also more than a drill ground for power mowers, or a guard against combine rustlers; it is a symbol of openness to others and respect for them.

The new machine and equipment sheds look simpler, even better kept than the older buildings did. There are not only fewer buildings, but they are of neater and simpler shapes. The round bins, often starkly white like sheds and farmsteads (there are few red barns here) and the occasional farm elevator, with its wire bracing, take on the look of abstract elements. Frank Lloyd Wright thought that the prairie house should be low, with deep overhangs, and should nestle into the land. The cornbelt cube, stark, high and blocky, eaveless, is its antithesis, a bold imposition on the land, a clear expression of human artifice.

This abstract regularity reinforces the dominant organization of the mile square grid. The intersections are marked with bright green signs—1400E, 800N—that tell you just where you are. Mailboxes read "The Bowers—Virgil, Doreen, and Lurleen: 1100N, 633E," a logical but curious way of anchoring the globe to the southwestern corner of your county. The row crops, too, follow the grid; driving the county roads in the late summer is like speeding through eye-level corduroy. The smaller country cemetery, located on a rise, never succumbed to the curvilinear cemetery planning fashion; its plots and marker stones are lined up so that even the dead are settled on the grid system. With the disappearance of windmills and trees, a new set of verticals, the phone and power poles, provide a regular, repetitive subdivision to that grid. The hedgerows that are left are now simple lines of trees; they read not as low, wide, rounded masses but as thin lacy walls, tracing field patterns and reinforcing further subdivision of the grid. Seen inscribed against a winter sunset, they seem to symbolize nature not only receiving the grid but becoming the grid.

That grid, surely the most extensive visible abstraction ever laid upon the globe, impresses most from the air. On a transcontinental flight one first sees it appear tentatively over Michigan, or maybe Indiana. Over Illinois and Iowa it *is* the landscape, with only the briefest of interruptions for steep terrain. By the great plains it has loosened; its grain is coarser now, because roads every mile are not economical, but the pattern remains. By the Front Range it disappears, only to reoccur in irrigated valleys. To understand it best, fly Illinois and Iowa in a small plane, at say 3000 feet above the ground. To do so is to see the ultimate expression of Cartesian rationality and Jefferson democracy, a noncentrist, nonhierarchical organization. It is an organization so powerful that even the few departures seem to reinforce it: the wrinkling of the land in Iowa like a topologist's diagram or the subservience of the interstate gracefully curving on top of it only to return to the half-section line, where land taking is shared and the township road system undisturbed. The buildings, simple, abstract cubes set square to the road, the circular counterpoint of bins or an occasional irrigation circle, look like pieces on a gigantic monopoly board, based on the rules of commodity and political equality.

It is a visual landscape that contemporary high-technology farming has molded but not disrupted. The disruptions are the landscapes brought by new residents, the curving roads, and artificial lakes of the wealthier exurbanites seeking that original niche of human occupation, the wooded stream valley, or the one side of a quarter-section, small-lot subdivision of strip septic tank suburbia, or the single lot with its long, low, dark-stained ranch house set on a "welcoming" diagonal. The farmer has a clear and bold vision of the landscape; the newcomers do not.

In many parts of the world the farmer is considered a sly, dimwitted peasant, but his landscapes are viewed as charming. In America the yeoman farmer, if not an image of sophistication, is still a central figure in the democratic dream, a figure enshrined by Jefferson and as much enriched by populism, grange, and dust bowl as he has been simplified by nostalgia and Norman Rockwell. His landscape is equally worth our understanding and our admiration.

Two Hundred Years of a Farm Landscape

LEWIS D. HOPKINS

Riley, as a landscape design intellectual, speculated on what he saw as a new rural landscape disrupting previous images of countryside.[1] I have lived the last fifty years a couple of blocks away from where Riley did, surrounded by the east central Illinois landscape of flat land, corn, and soybeans, the landscape that most directly framed his speculations and questions. I have also driven hundreds of times beyond the eastern edge of the cornbelt to a family farm in north central Ohio that has been in my family for 180 years. Thus, the way in which I see and know this Illinois landscape is different from the way in which I know and see the Ohio landscape. I use this contrast to suggest that the frames we set up to interpret the landscape constrain and enable the kinds of interpretations that can emerge.

In Riley's framing, simplified, the new rural landscape results from an evolved landscape of farming being confronted by newcomers who are merely residents. His derivation of this frame is evident in the questions he poses. What do you see? Why is it there? What do these newcomers seek? He starts with an observed landscape, then claims that past explanations no longer apply, and finally focuses on what the newcomers, the subset of people given agency, want. Although he poses many questions, I infer that many are rhetorical. In conversation, he would have agreed quickly that there are regional differences. I think he would acknowledge that the framing is specific to places like east central Illinois in which the landscape pattern is easily delineated, easily seen and categorized. I am not aware that anyone has pursued research from his frame, and my purpose here is neither to test it nor to contest its details.

As an environmental planning academic and an in-the-dirt participant in creating a farm landscape, I use a different frame in a different kind of landscape. In simplified form, my frame does not assert a new landscape, and it roughly reverses the questions: Who was and is there? What are they doing? What landscape emerges? This frame leads to different questions and enables different kinds of explanations. A proper research design would apply each frame to each place. I do not have intimate knowledge acquired over two hundred years to apply my frame in Illinois, and Riley's frame breaks down quickly in my Ohio neighborhood because the landscape is not easily delineated and the newcomers are not disruptions there.

My identification with a particular farm and farm neighborhood, although unique in its particulars, is an instance of a broader phenomenon as evidenced in the stories in Hemalata Dandekar's *Michigan Family Farms and Farm Buildings: Landscapes of the Heart and Mind.*[2] She interviewed multiple generations of farm families that had held their farms for more than one hundred years. We hear in their stories how farms came to be as European settlers arrived and what decisions families made about investments in immobile and mobile equipment, on and off farm jobs, temporary migration for jobs or educations, diversifications across risks among family members or locations, all in tandem with roots growing in a particular place. It is these kinds of stories and a design idea, "historical depth," that helps make sense of our Ohio neighborhood. Though his focus was usually cities, I associate historical depth with Kevin Lynch.

> In central, public places, we would think less of freezing some idealized moment than of allowing the scars of time to accumulate visibly. By selective collage we would try to make everyone aware of the depth of time and its cross-current. Fragments of former structures would routinely be incorporated in new construction. Current use would be asked to signify previous use, or the activity from which it evolved. Conflicting views of the past should be exposed, traces be erased and redrawn as our views of the past shifted. In other words, in place of sanitizing the past by purifying and freezing it, we should contaminate the present with it.[3]

A brief description shows why starting from what we see would be so much more difficult in our Ohio neighborhood than in Illinois. Our neighborhood is at the headwaters of a creek where the glaciated edge of the Appalachian plateau crosses the continental divide between Lake Erie and the Mississippi basin. Driving in our neighborhood, within three miles around our farm, we get a rural perspective on how the world works. A gap in a floodplain grain field reveals where a pipeline went through to handle the recent surge in natural gas production. A small pallet factory is tucked back in some woods. Metal-clad sheds six hundred feet long house chickens or hogs. A modern "sugar house" boils maple sap. An Amish workshop fabricates sheet metal. A recently fracked gas well has been capped as not sufficiently productive; an oil well in the neighborhood for at least fifty years may be silent when the price of oil falls. Clusters of ranch houses are close to the road. The organic dairy farms control grazing cows with movable electric fences to intensively manage pastures. Bright white PVC fence posts hold four strands of nearly invisible wire. The Amish farms retain the farming practices of one hundred years ago: horse-drawn plows, loose hay, shocks of grain in fields, and stationary threshing. There are also "English" medium-size modern dairy operations and beef cattle farms. Rural business operations include used equipment sales, sawmills, mechanics, machine shops, and building supplies. And within three miles there are a state nature preserve and a regional landfill.

I recently tracked down again the carcass of a Model T Ford that has been rusting into the ground in the middle of our woods. I do not know who put it there or when. Most likely it arrived there in the 1930s when a pre-1920 Model T would have been worthless, obsolete technologically. It was hidden in the woods, unlike the "Camaro in the Pasture" displayed for sale near the road in Riley's description of a new rural landscape. It is an instance not only of rapid technological innovation and obsolescence, but also of disposal. Seventy years ago, along the then gravel road along the creek there were "No Dumping" signs fronting small ravines filled with trash. Today, we drive to the regional landfill, where we get free access as negotiated compensation because it is in our township. We can see the landfill mountain from our farm. We also walk the road ditches picking up litter. We burn bonfires of fallen branches, invasive plants, and scrap wood. Disposal is part of the rural landscape. Its visibility, whether in the woods, in the ditch, or as a landfill mountain, is a valuable reminder that the stuff is going somewhere. Driving the rural landscape is a way to learn not only where our food, lumber, oil, and water come from but also where our trash goes.

On our "Home Farm," as shown in the aerial photos here, the lane follows a small creek to the center of our eighty acres. The terrain is just sufficiently incised, about 25 feet, to create significant variation within its 80 acres. This farm landscape has been happening for two

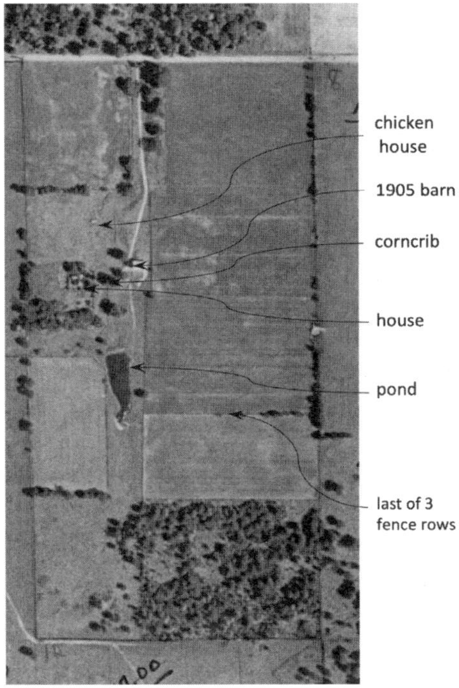

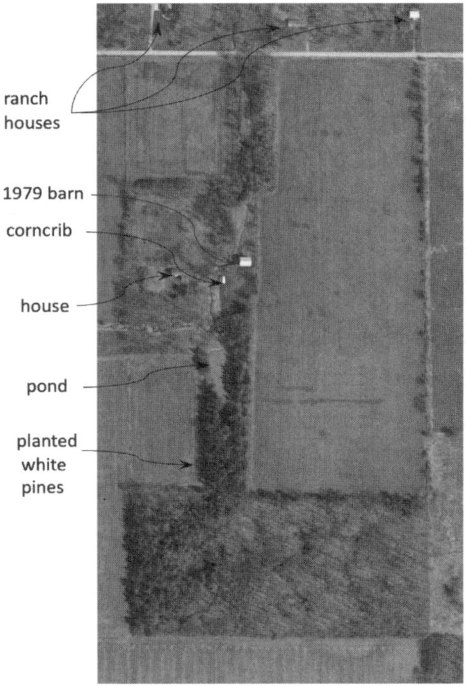

left Aerial photograph, 1958. (Hopkins family collection)

right Aerial photograph, 2006. (Hopkins family collection)

hundred years, as summarized in the accompanying table. My great-great-grandfather and a brother bought the eighty-acre farm in 1842. In 1844 my great-great-grandfather married a neighbor's daughter. After he died in 1851, she relied on siblings on nearby farms. Most of the land was wooded, which required clearance for crops and pasture but provided fuel, lumber, game, and nuts. Our farm was well suited for these nineteenth-century practices. The creek created pasture terrain with access to water, and springs at the base of the hills provided water for both house and barn.

In 1876 the second son, James, my great-grandfather, married a neighbor, Huldah, whose family had accumulated at least 320 acres, built a large house, and purchased a piano for her eighteenth birthday. When her uncle died in 1898, the auction poster listed 410 sheep and 100 chickens. The new couple lived on eighty acres about a mile away from our Home Farm. James and Huldah not only raised sheep but also ran an apple cider mill and made apple

200-Plus Years of a Family Farm

PEOPLE	DATE	HOME FARM	NEIGHBORHOOD
Native Americans; then confiscation by European settlers	Prior to 1820	Wooded creek headwaters incising glacial till	Wooded creek headwaters incising glacial till at edge of Appalachian Plateau European settlers after 1820
Brothers **William** and Andrew **Patterson** purchase 80 acres.	1842	Clearing fields and pasture; grazing sheep and subsistence farming	Scots and Scots-Irish neighbors developing similar farms, typically 80 acres; grazing sheep, apple orchards, subsistence farming
William marries neighbor **Anna**: William, **James**, John	1844		
William dies leaving **Anna** with lifetime estate in farm	1851		Anna's siblings and in-laws are on adjacent farms, probably farming collaboratively
	1867	New house built in part on old foundation	
James marries neighbor **Huldah: Ethel,** four others	1876	Apple orchard 2 acres; woods 10–20 acres; oats, wheat, hay, corn in 4- to 8-acre fields; sheep in pasture; chickens, hogs, milk cows for own consumption	**James** and **Huldah** on 80 acres a mile to northwest among **Huldah's** family's farms: Cider mill, 145 sheep
	1901		**James** and **Huldah** build large barn
Ethel marries **Clayton Hopkins:** Ernest, Esther, **Dean,** Martha	1903	Apple orchard 2 acres; woods 10 acres; oats, wheat, hay, corn in 4- to 8-acre fields; sheep; chickens, hogs, milk cows mostly for own consumption	Small store at an old mill site a mile east
Live on Home Farm with **Anna,** who dies in 1904	1905	New bank barn	
Ethel and **Clayton** move family to town for high school	1919	70 sheep; chickens, 15 hogs, 7 milk cows, sold at auction. Take 2 cows and 2 horses to town.Farm rented; wooded areas converted to pasture except one corner	Hay hauled to railhead 7 miles to ship to Cleveland to feed horses Other farms from **Huldah's** ancestors sold outside family
Ethel and **Clayton** return to Home Farm	1935	Walnut trees planted on part of previous garden south of house; sheep, chickens	
	1951	Pond built; Sell eggs to general store three miles away.	
Dean buys siblings' shares of Home Farm at deaths of **Ethel** and **Clayton**	1963	Plant white pines, cypress, green ash, sweet gum in south pasture and infilling woods. Rent fields and pastures to neighbors.	Multi-farm dairy operation expands to seven farms
	1973	New, noncommercial orchard, 47 trees	First ranch houses of farming families along road.
	1979	New barn replaces burned 1905 barn	
	1986		Multi-farm dairy operation goes bankrupt.
Dean dies; Home Farm transfers into a trust for his descendants	1993	Tree planting and hobby farming continue. Fields and barn rented to dairy farming neighbors.	Amish and buggy Mennonites buy bankrupt dairy farms; continue to purchase parcels as they become available;
	2017	Fields in intensive grazing for organic dairy operation of Mennonite neighbors	Ranch houses of non-farming residents increase along roads; dispersed rural businesses.

butter. The auction poster for the sale after his death in 1914 includes "145 head of sheep" and "100 half gallon jars" that were for apple butter. They were not reliant only on farming. In 1901, James and Huldah built the most massive and elegant new barn in the neighborhood, which included a sunburst design in one of the gables. During the barn building, their daughter, my grandmother Ethel, met my grandfather Clayton, who was in the barn-building crew. He was in one sense a newcomer and nonfarmer, as he grew up in a town seven miles away. He "became" a farmer.

Ethel and Clayton married in 1903 and moved onto our Home Farm, taking care of my great-great-grandmother until she died a year later. The Home Farm was later transferred to Clayton and Ethel for one dollar. Such details of transfers are relevant to understanding the rural landscape because they help explain who is there, then and now, and how wealth and status accumulated: Neighbors married; descendants bought out their siblings' shares of their ancestral farm. Mixed-age extended families sustained farming operations as a neighborhood and acquired neighboring farms that became available. These patterns continue in our neighborhood today.

In 1905, Clayton and Ethel built a bank barn similar to, though not as grand as, the barn through which they met. The farm landscape continued the pattern of large, tall barns designed for storing loose hay and stationary threshing of grain to yield a straw stack out back (see photo page 106). City cousins visited and played at being real participants in farm activities. In 1919, my grandparents moved their family to a city thirty miles away so that their children could go to high school. This decision again evidences the ambiguity of farmer or newcomer, from town or from farm. They rented the farm outside the family or loaned it to relatives until they returned in 1935. This choice was feasible because they had inherited the farm without having to pay for it and because my grandfather had grown up in a family of building contractors, so he knew how to make a living in town. They did not rely on the farm for income. The house in town had a small barn, as many did because automobiles were still uncommon. They took two horses, two cows, and two turkeys. They had a large garden and a small orchard. My father delivered extra milk to neighbors.

In 1935, their children educated and the Depression limiting opportunities in town, my grandparents returned to the farm. Forty sheep were shorn as recently as 1944, continuing practices brought from Scotland one hundred years earlier. In the 1950s, my siblings and I were the city folks visiting our grandparents' farm, helping to collect eggs that were sold to

left Pond looking southwest (ca. 1953). (Hopkins family collection)

right Pond looking south, surrounded by pine trees planted in 1960s (ca. 2018). (Hopkins family collection)

the general store three miles away. My grandparents were living on land rental, egg money, and Social Security. We fished with long bamboo poles in a pond built to Soil Conservation Service standards in 1953, while a neighbor's heifers grazed around us. The pastures, fields, fence rows, and a gradually thinning woods remained. The photographs above show this landscape.

As early as 1944, my father wrote a letter to his parents: He was becoming a successful lawyer and thought he could eventually "keep the farm in the family." This was important to him and to his mother, both of whom identified strongly with the farm as place and as family heritage. His father, on the other hand, thought it folly to hang on to this farm, well suited to the nineteenth century but not to the twentieth. It was no surprise that my father bought his siblings' shares when my grandparents died in the 1960s.

For my father, the farm was a place to emulate in his retirement his own life experience and share it with his children and grandchildren. The pattern of renting the tillable ground, barn, and pasture continued and continues to this day. We are not reliant on it for a living, which means we do not have to scale up or intensify production. We gradually transformed most of the former pasture to trees. Following a succession of changing forester advice, we

left The 1905 barn in snow with straw stack; footbridge without railing (ca. 1940). (Hopkins family collection)

right The 1979 barn with creek in flood over footbridge with railing (ca. 2015). (Hopkins family collection)

planted white pine, green ash, sweetgum, cypress, walnut, red oak, white oak, tulip poplar, and sugar maple. On their own initiative, elms, walnuts, willows, and osage orange reclaimed parts of old pastures along the lane and creek. We planted a new, noncommercial and non-subsistence orchard north of the house where a large orchard had been from at least the 1880s to the 1920s.

In 1979, the bank barn built in 1905 burned. This event was a symbolic marker in the landscape's evolution. By that time, a childhood playmate of my father's owned at least seven dairy operations on farms surrounding us. Tall bank barns were replaced with low barns and milking parlors. More fields were used as pasture. Fencerows disappeared as the crop rotation shortened to hay and corn and machinery grew larger. Fence posts were treated wood. Our family discussions acknowledged that one way to honor the 1905 barn, which was built to the best practices of its day, would be to build a barn fitted to the new farming landscape. We decided, however, to replace our tall bank barn with what I think of as a replica: a bank barn, but smaller and with horizontal trusses instead of the vertical trusses best for handling the loose hay of 1905. There were enough large beech trees in the woods, already growing long before 1842, to provide posts and beams for the new barn's ground floor. It was designed so that it would still work for heifers from a modern dairy operation—a functioning replica, not a mere façade.

By 1990, the landscape was going through a time warp. The multifarm dairy operation surrounding us went bankrupt. The ranch house built in the 1970s as its headquarters seeded a cluster of ranch houses still occupied by the owner's descendants. The bankrupted farms were bought by descendants of other century-old neighborhood families and by Amish and buggy Mennonite families new to the neighborhood. These newcomers have brought capital from elsewhere to sustain or recreate the continuing farm landscape of dispersed entrepreneurs making deliberate choices about modern technologies. The old-timers survive by adopting the newest technologies or by living on their ancestors' land without living from the land. Most of the occupants of ranch house parcels around our block are descendants of long-time families or the next generation of neighborhood Mennonite families unable to establish an operation of sufficient size to make a living entirely from farming.

Organic milk production using modern technologies yields a higher price in an otherwise declining market for milk. The bright white PVC fence posts meet the organic standards that the treated posts of the 1960s do not. The Amish use the osage and locust posts that they find in their woodlots and in ours. The ranch houses almost all have extra sheds for a hobby or memory of some sort, such as animals or a woodshop. Some sheds are livelihood businesses making windows, repairing machinery, or running a sawmill. Some farms are raising vegetables or flowers. Some farmsteads have been turned into stores selling hardware, farm supplies, or construction materials. The ambiguities of town and farm, farmer and newcomer are still apparent in this rural landscape.

We maintain the picket fence, though its only function is to sustain memory and skills. We battle invasive garlic mustard, honeysuckle, oriental bittersweet, and floribunda rose. We transplant wildflowers from the never-tilled woods to newly wooded areas from which the spring ephemerals have disappeared from the seedbank, and we add missing species appropriate to the place. A few white and bur oak seedlings are positioned carefully in old pasture where we can control their surroundings in order to create open-grown, spreading trees for one hundred years from now. All the huge open-grown oaks that survived for 150 years after European settlement have fallen. When a white pine that we planted in the 1960s is blown down, a neighbor's sawmill can turn it into boards. When the ash trees were dying from emerald ash borer, we sold them to Amish neighbors who horse-logged them. When we want furniture-quality boards from walnut, an "English" neighbor can saw and kiln dry. We planted a few of the latest resistant American chestnut seedlings. We plant apple trees as those planted fifty years ago fail, and in fall we haul apples to a local cider mill, farther away and smaller than my great-grandfather's mill. We host our grandchildren. This 2006 landscape is shown in some of the photographs here.

The sunburst decoration from my great-grandfather's 1901 barn a couple of miles away is now on the gable of our Home Farm corn crib. The current owner of that farm salvaged the sunburst when that barn collapsed a few years ago. Recently, we found a neighbor to rehab and tune Great-Grandmother's piano, passed on to my grandmother and too heavy to be stolen in burglaries that have taken other historical items over the years.

This account is primarily an anecdote. I have no intention of claiming intellectual insights or generalizations from its particulars. It suggests, however, that at least in some places following people over time might lead to different understandings of an observed landscape than asking newcomers what they are seeking when disrupting an observed landscape in a highly evolved state. In my Ohio neighborhood there is too much variety in the physical landscape and in the reasons the people are still or just now here.

Our farm is now in a trust for my father's descendants. Even among my generation, much less the next two generations, there are different interests, ideas, commitments, and visions of why we are there. Some of us want it to be a farm, a place that keeps us in close touch with commercial agricultural production. For some, the farm is primarily a place to be with family. Others are concerned that it generate enough income to cover expenses. For some, it presents an opportunity to show the next two generations a modern milking parlor or how to fell trees, haul apples to a cider mill, or just to have some clue that a rural neighborhood exists. Our neighbors, regardless of how many generations or the portion of their income from the land, also seek a wide variety of different things. If there is anything in common, it is the opportunity to do something that they could not do "in town."

This landscape obviously matters to me. Our eighty acres is much more engaging in its variety than a huge metal shed with two huge machines at the edge of a square mile of corn or soybeans on flat land. As an academic largely separated from the dirt but now retired, I can actually build sediment control structures, confront invasives, plant trees, experiment with restoration agriculture, and participate in decisions about herbicides and pesticides.

Does it matter that the land has been in my family for 180 years? Yes. I am constantly reminded that I come from settlers who took land from an Indigenous population, that my status is built on generations of wealth and status accumulation, that my father was born at home in that bedroom, and that choices about career, family, place, and landscape propagate across generations. This past is part of my present because it contaminates what I see in the evolved landscape of the present.

NOTES

1. Robert B. Riley, "The Camaro in the Pasture," in *The Camaro in the Pasture: Speculations on the Cultural Landscape of America* (Charlottesville: University of Virginia Press, 2015), 92–99.

2. Hemalata C. Dandekar, *Michigan Family Farms and Farm Buildings: Landscapes of the Heart and Mind* (Ann Arbor: University of Michigan Press, 2010).

3. Kevin Lynch, "The Image of Time and Place in Environmental Design," in *City Sense and City Design: Writings and Projects of Kevin Lynch*, ed. Tridib Banerjee and David Southworth (Cambridge, MA: MIT Press, 1990), 631.

Snapshots and Fragments

Manitoba Farmstead Shelterbelts and Their Stories

BRENDA J. BROWN

Drawings by Emma Dicks, Gel Ilagan, Michaela Peyson, and Brenda J. Brown

A scholarly comparison of the actuality of farm landscaping with the advice of designers and improvers, however, might show that farm families have evolved their own images of "farmstead" independent of professional fashions.

—ROBERT B. RILEY, "Square to the Road, Hogs to the East," 1985

Farmstead shelterbelts' layered rows of trees can be seen simply as planted members of the family of stark, clean, vertical forms punctuating North American prairies and plains. Viewed from outside and afar they might suggest medieval fortresses, and their tall conifers are reminiscent of cypresses enclosing an imperial garden. And indeed, these trees create "rooms," or gardens—landscapes within the larger landscape, spatially and microclimatically different, and signaling protection from, the surrounding wind-swept expanses. I have traveled across prairie and plain, and they have long intrigued me, perhaps partly because until recently I usually viewed them from outside—from the road or from the air.

Yet my route to this research has been oblique. In 2018, as artist-in-residence at Riding Mountain National Park, I began work on a video project: *Wind in Trees/Trees in Wind*. Thinking about trees, wind, and cultures led me to conceive a companion video project on Manitoba shelterbelts.[1] Considering further, I realized that a study of the people-tree relationships inherent to farmstead shelterbelts could be a window onto Manitoba's changing rural landscapes and the experiences and stories of those who live in them. And so the sound/video project expanded to also encompass elements that will follow here: interviews and transcriptions, tree measurements and identifications, photographs, diagrams and other hand-drawn documentations.

In summer 2019, I mentioned to Bob Riley where my concerns with wind and trees seemed to be leading me, but at that point my idea of the work's shape was nascent and any comment Bob may have made was minimal. Only after I began to imagine how text and images might be organized did I realize some of the ways it reflected Bob's influence—influence beyond his own work on rural Midwestern landscapes. As you will see, this piece combines interviewees' stories in their own words, photographs of them within their landscapes, and annotated plan, axonometric, and sectional drawings of farmsteads with their shelterbelts. As others have noted, one of the things many students appreciated about Riley's classes was the wide-ranging images and readings we considered so as to better comprehend landscapes' ubiquity and power. After conceiving this essay's format, I remembered the stories in Ronald Blythe's *Akenfield*, and Richard Westmacott's drawings of African American gardens, both of which Bob used in his seminars. I remembered too, how along with images of landscapes shaped by people and marked with their traces, Bob showed photographs of landscapes with their inhabitants. And I remembered my own discovery during those years of Paul Strand's books in which images of people and images of landscape made a whole portrait of place and how I shared these with Bob.

I do not know what Bob would make of this work and my presentation of it. It is not the scholarly comparison he suggested in 1985. Still, I like to think that he would get a kick out of it.

These are preliminary findings about Manitoba shelterbelts—farmyard shelterbelts in particular—and their significance for families who live among them. I hypothesize that Manitoba's shelterbelts are not only significant and distinctive visual and ecological elements of the rural landscape but that they are also psychologically, socially, and culturally meaningful. Participants became involved in this research because they were interested when they learned about it from friends or conservation district managers, thus I do not claim it is a representative sample of all who live within farmstead shelterbelts.

The interviewees' stories included are fragments of much longer accounts. As this is written, twelve properties in different regions of the province have been documented and

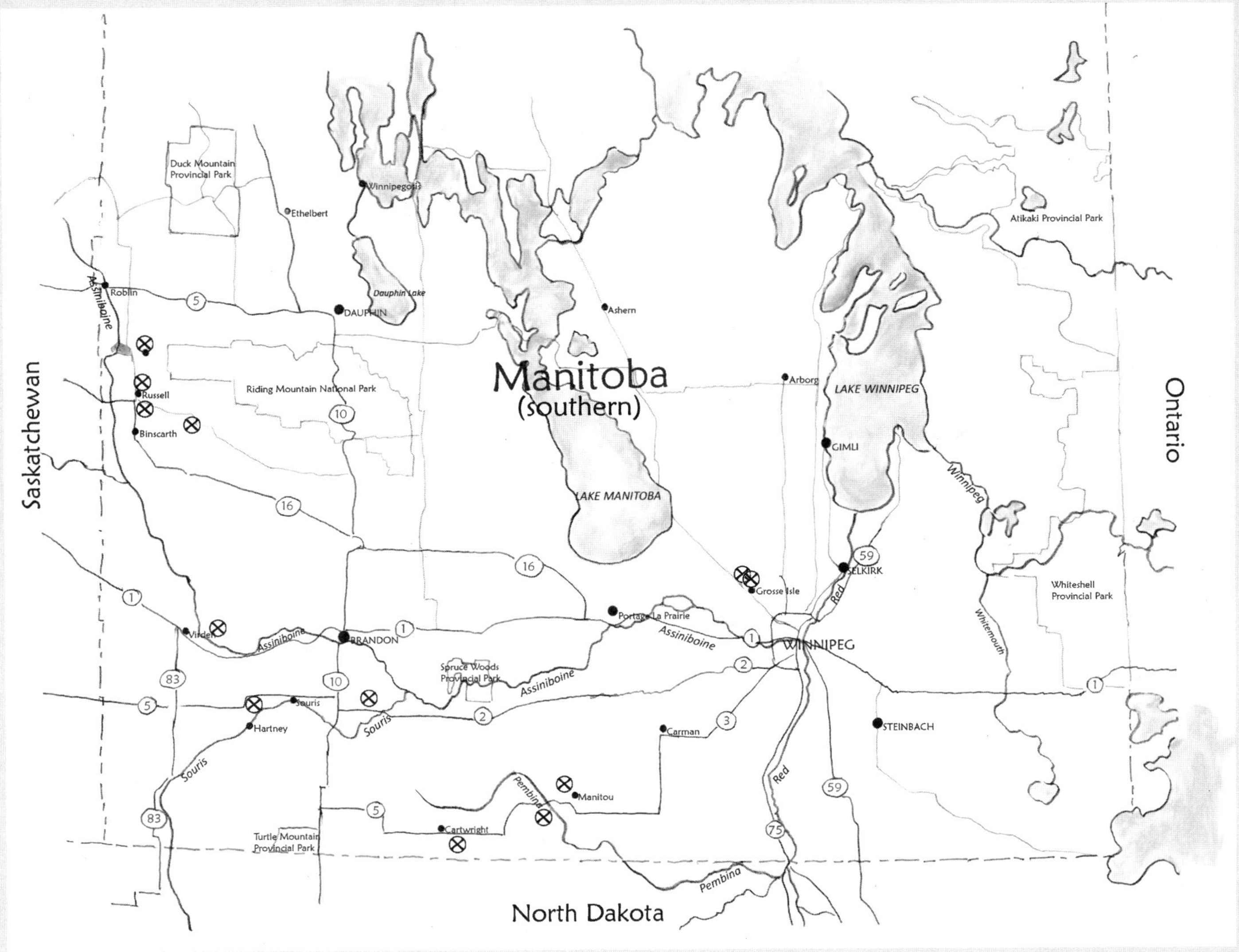

Map of southern Manitoba, indicating farmstead sites visited and documented in 2020.

their occupants interviewed (see map). While these are dispersed fairly widely, approximately fifteen more properties in other areas—most obviously the southern Red River Valley and the Swan and Whitemud watersheds—will be investigated and documented in the coming season. None of Manitoba's Hutterite or Mennonite farms or their occupants, so significant to Manitoba's agriculture and landscapes, have been documented yet either, and I hope this can be corrected.[2] Due to limited space, only seven properties (and their owners) are featured here. They suggest the diverse locations, property sizes, shelterbelt species, individuals and families thus far encountered.

Manitoba's farmstead shelterbelts are not purely vernacular expressions; they have been shaped, aided, and supplied by government programs and agencies. Early European settlers made use of treed areas along rivers, initially, and most notably, along the Red and Assiniboine.[3] Serious and extensive agricultural development really began following Manitoba's incarnation as a Canadian province and the subsequent 1872 Dominions Land Act that led to a township and range survey system. Similar, though not identical to, and quite deliberately derived from the United States' Jeffersonian grid, its basic settlement units were 160-acre homesteads.[4] Settlers were sought from southern Ontario and the United States as well from Europe.[5] Lore has it that many of these settlers planted trees around their homes for wind protection and beauty, but also because they wanted to live in places resembling those from where they came—and trees were part of that.[6] However, their planting efforts were frequently unsuccessful as the species, stock, or seed they brought or imported often did not travel well and was ill-suited to Manitoba prairie soils and climate. Reflecting the Canadian government's vested interest in populating the prairies, the Forestry Nursery Station in Indian Head, Saskatchewan, was established in 1901, followed not long after by one in Brandon, Manitoba. Their purpose was to propagate and supply farmers with free plants hardy enough for Canadian prairie conditions.[7]

While the nurseries and programs changed names and government departments over the coming years, federal support for shelterbelts continued until 2013, some 112 years. Prior to the 1930s, trees were used primarily to shelter farmyards; however, with the Dust Bowl and the respondent 1935 passage of the Prairie Farm Rehabilitation Act, shelterbelts were promoted to counter soil erosion and drought in fields. Capacity increased accordingly at the Indian Head Nursery, now part of Canada's Department of Agriculture's Dominion Experimental Farms Branch, and workers there assisted in the design and promotion of field shelterbelts through the Prairie Shelterbelt Program (PSP). In the 1960s, and until its demise in 2013, federal support for field shelterbelts was part of the Prairie Farm Rehabilitation Administration, the PFRA, to which many interviewees refer (no date). Between 1937 and 2017 over forty-three thousand kilometers of field shelterbelts were planted in Canada's prairies.[8]

Farmers ordered the seedlings they wanted and were responsible for planting, maintenance, and sometimes shipping, but the seedlings themselves were free. While work at Indian Head included research and development on different tree species, varieties, and hybrids, it also provided farmers with a particular palette, though supplies could vary from year to year. Many of the farmstead shelterbelts studied have been built with PFRA trees. This partly explains the similar species used at different properties, but it also reflects the species that have over the years proved adaptable to the region's conditions.[9]

While shelterbelt creation and care continues to be among staffs' responsibilities at some Manitoba Soil Conservation Districts, withdrawal of federal government support has combined with other factors to reduce creation and maintenance of Manitoba shelterbelts. Plants that farmers used to get free from the national nursery in Indian Head or the province's Pineland Forest Nursery must now be purchased from private vendors. Changes in tillage practices, crop processors' shifting preferences, larger machinery, larger acreages, fewer farmers raising cattle, and related economic concerns have also played a role in old shelterbelts' disappearance and the dearth of new ones. Farmstead shelterbelts, like field shelterbelts, are disappearing. As a farmer in the Pembina Valley observes, "When I look around and I see a farmyard site, it represents people living there. But that isn't always the case. And oftentimes, they sit vacant for a while and then the trees are taken out as a bigger operation comes in. That's always sad. So then the countryside seems less populated by people."

However, farmstead shelterbelts' disappearance is slower than those of fields, and as will be shown, they are also being maintained, added to, and created. Shelterbelts around farmsteads have been shown to be particularly valuable for energy conservation, reducing home energy costs, improving outdoor work conditions, reducing equipment and structural maintenance expenses, and decreasing the size of snow deposits and wind effects on roads.[10] Observations and anecdotal reports are bolstered by Ashton and Richards's study (2014) suggesting that farm families value the shelterbelts around their farmsteads above those elsewhere on their lands.[11]

Although I cannot claim this is a representative sample of all who live within farmstead shelterbelts, it is perhaps representative of those who live with, and care about and for them. They range from first-generation to sixth-generation Canadians. Many, though not all, have family connections to their homesteads or at least to the local area in which they reside. One farmer farms over three thousand acres; another's farming is limited to a fenced family garden within the homestead. The scale of the shelterbelts themselves also varies considerably as is reflected in the three different scales used in the drawn plans of the farmsteads on the pages that follow. A few interviewees are retired or semi-retired; some farm on the side, others are fully engaged as their farm is their livelihood. One site's shelterbelt is actually derived from existing oak-rich bush; another's is the result of forty-plus years of a couple's labor. The shelterbelts give the farmsteads different shapes; some encompass only the house, others many of the utilitarian outbuildings as well.

The tree rows of prairie farm shelterbelts were originally planted primarily for wind and soil protection; however, their additional value for carbon sequestration, microclimate creation, livestock protection, and wildlife habitat has drawn increasing attention. The people with whom I have talked speak about all these things; they speak about their shelterbelts as things alive, cared about, and cared for—and they speak about more besides.

Henning

My parents lived here from 1978 till 2006, and then my wife and I moved here in 2006. My wife and I and our three sons live here now; they're nine, eleven, and thirteen.

My parents immigrated here from Germany in 1977 to start farming. My dad wanted to farm. He grew up on a farm in eastern Germany, but then after the war, it got separated because the East and West. . . . They were in the West and had an opportunity to come over here and look for a farm and so he rented the farm. And yeah, that's basically history. They started in 1977 in a different yard site and then moved here in '78, '79 and the house wasn't finished, but they finished it and moved in.

I grew up here. I was born in '79. And I was here till until I went to university in '97. I studied agriculture at the U of M.

We just farm grain crops: Wheat, canola, oats, barley. We've grown anything from flax to soybeans to lentils to peas to Canary seed.

CS—*Cornus sericea* / FP—*Fraxinus pennsylvanica* / PP—*Picea pungens* / PV—*Prunus Virginiana melanocarpa* / SA—*Salix alba* or *Salix acuifolia* / SV—*Syringa villosa*

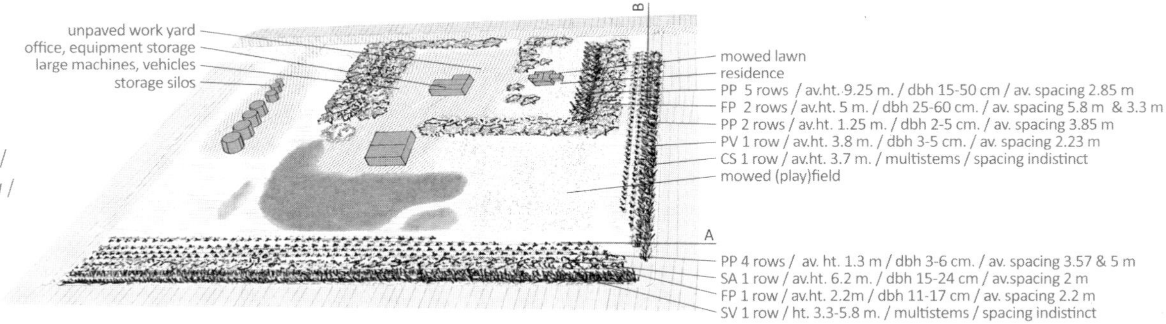

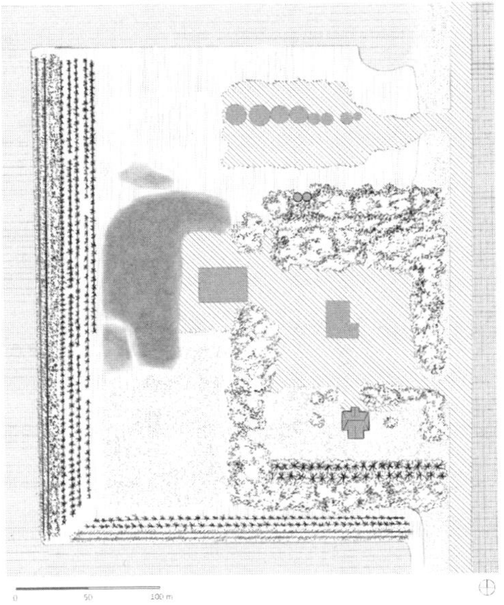

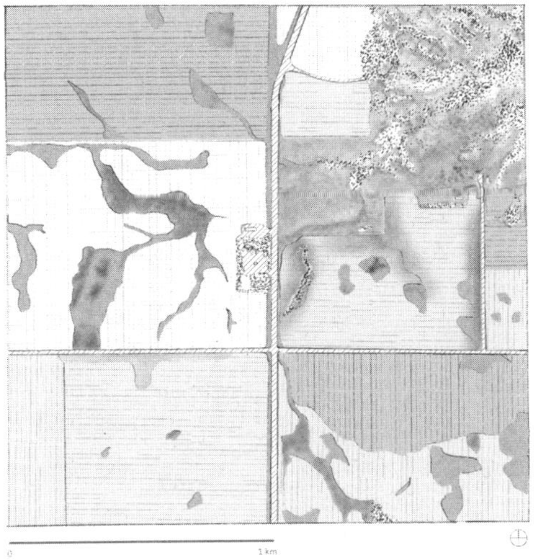

The economics of wheat is mostly for rotation; oats is probably our top money-maker. Canola is one of our most productive crops as well, and barley is this year because of the late spring we had to put something in. Soybeans have been great for rotation but haven't been great economically. So we've switched away from them this year, taking a break.

The original shelterbelt was here when my parents got here; then in 2014, I believe or '15, we added the extra one outside the yard.

We're putting in another row out here on that side of the trees. There's four rows on that side. And then on the west side, we have I think six or seven rows of trees out there. And then we have four rows on the north side now. We put those in about five or six years ago because we wanted to expand our yard for the farming operation and didn't want to be out in the middle of nowhere. So we moved all our bins out there—most of them—and started our grain-drying system out there, added a shed there. As much work as it is, we do like the trees for around the yard and hopefully, maybe one of our boys will take over the farm and they'll be big and tall by then—those trees.

I don't like the amount of work involved in keeping them weed free or as weed free as possible. I wish there was a silver bullet that you could, you know, help with that, but there doesn't seem to be any one chemical that you could spray or—I thought about putting grass out there, but haven't done that yet. Just to keep the weeds down and just mow instead of till. But so far we've just been tilling. We have what they call an eco-weeder, which is a rotary tiller that we purchased a few years back to help with that and some hand picking in the beginning was a lot.

We farm 3,300 acres. It's a good size and it's a comfortable size for one family so to speak, and we're happy with that size and not really looking to expand. We just like to maintain where we're at, and do a better job with those acres instead of trying to get extra acres and maybe not pay as much attention to those acres as we should.

We were lucky this year we had lots of timely rains. So [the trees planted] seem to have established fairly well. The first year that we put them in we ended up watering a lot because we weren't getting great rains so we went around and watered a lot of the spruce, especially, that were struggling. I would say we're probably at 85 to 90 percent survival rate on the spruce, which is all right I think.

I don't think there's anything unusual about how we use the shelterbelts around the farmstead . . . mostly for shelter from wind and snow and a little bit of shade in spots . . .

Henning and south-side shelterbelts, 2020. (Photograph by Brenda J. Brown)

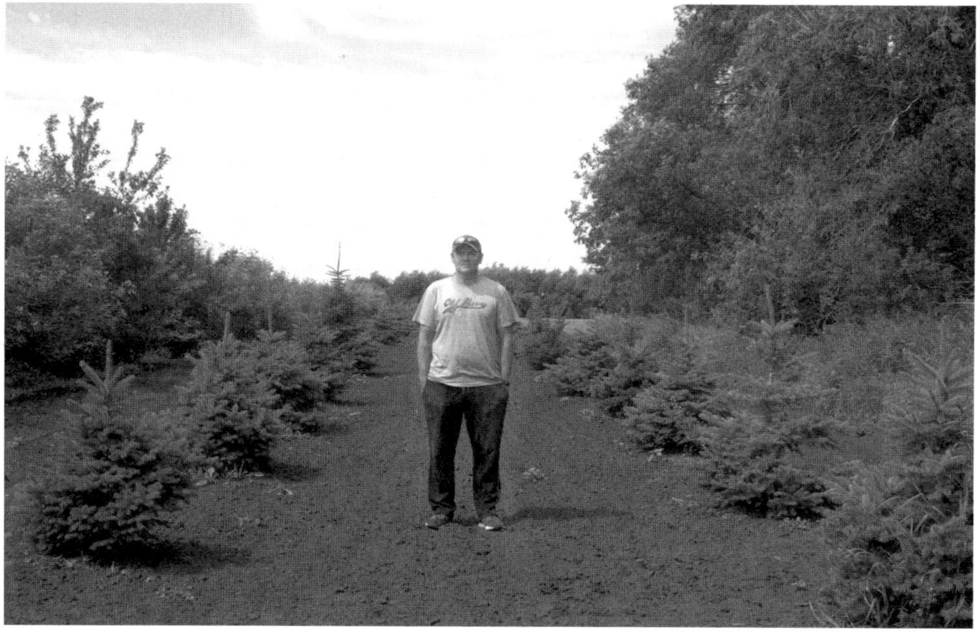

mostly for some privacy and to have a kind of enclosed area for the buildings. A lot of it was to keep as much truck traffic out to the north and not in the yard where the kids were playing when they were young. Now that they're quite a bit older, it's not near as bad, but that was a lot of the reasoning.

Andrea Gorda

We moved in December 6, 2007.

Five people live here: my husband, myself, and our three children: Katie's now thirteen, Jack is eleven turning twelve soon, and Madeline is eight.

Right here we own 9.96 acres. And then my husband shares one quarter with my father-in-law. And then my father-in-law has two more quarters. And on one of those quarters, we have some pasture that we have our cows on. The fields around us are owned by . . . various local families. They're not ours.

When we came and looked at [the property], we were surprised by how big the yard was, actually, because when you drive by you don't necessarily see the whole yard. . . . So when we moved here, the lilacs were here, the Scots pine were here, the Siberian elm were here and the big lilac at the end of the driveway. And same thing with the spruce and pine. The shelterbelt in the back and the willows were here as well. And most things were here. What we've added on are the Assiniboine poplar and some of the spruce. They were planted in 2008. And they come actually from Alberta. My mother-in-law was dying, so my sister-in-law came out from Alberta and her husband, my brother, or my husband's brother-in-law, brought these trees that his mom had given them, "Give these to Andrea and Glen." So he and I planted them. . . . In the back, far back kind of between the field, our property, and the canola field, there's a line of spruce there. And we got those from PFRA and we planted those on the eastern portion of the pasture. We did plant another row of ash and poplar and only a few of those have survived because the cows did get in and chewed them down. . . .

The kids will haul all manner of materials into the shelterbelts and then make forts in them. Like there's that little patch of caraganas and there's still stuff there. So they do that. Because these shelterbelts are dirt, there's been a number of attempts to dig to China. . . . When Jack was four—between four and six years old—he would dig holes like this either to China or looking for fossils or gold or whatever. And then in our shelterbelts too at some point, somebody had buried a large number of household articles. So every time you go through there it's like an archaeology, I think, 'cause the kids, sometimes if they're bored they'll go and look through the shelterbelts and pick up pieces of china and pieces of shoes and pieces of leather and glass and bottles—we've found a bunch of bottles and stuff—and they gather it up and then sort through it.

I think the farmstead shelterbelts are kind of an indicator . . . when you're driving down a road, and you see something that's obviously planted. It's like, 'oh, well, there's a yard site,' or 'oh, there's an old yard site.' So it gives you an idea of history . . . there's a huge number of farm sites around here that are abandoned, but you can still tell that they were farm sites by the tree species that are there and the formations in which they were planted. And sometimes, especially now with the hunger for more acres, they just get bulldozed. And it always, I don't know, it kind of breaks my heart that there is so much either pain or laughter, or Christmases, birthdays, new babies brought home, people died in those places, and there's

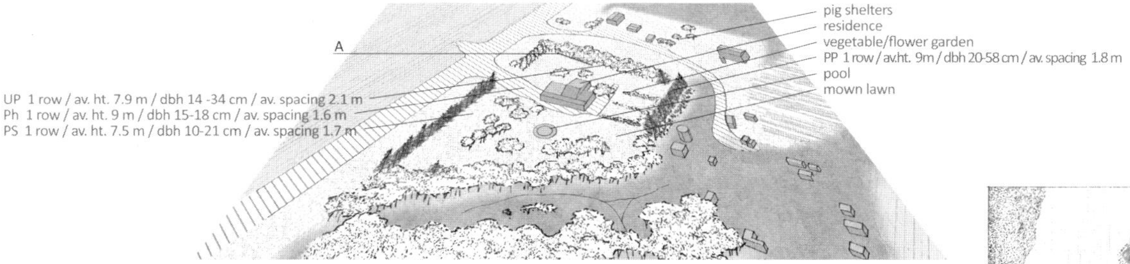

UP—*Ulmus pumila* / PP—*Picea pungens* /
PS—*Pinus sylvestris* / Ph—Poplar hybrid

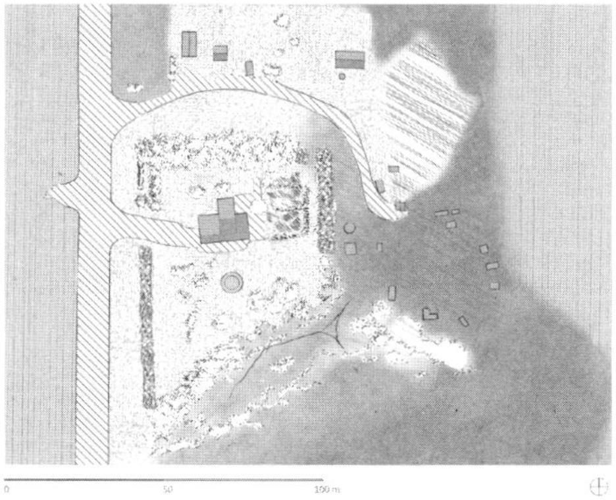

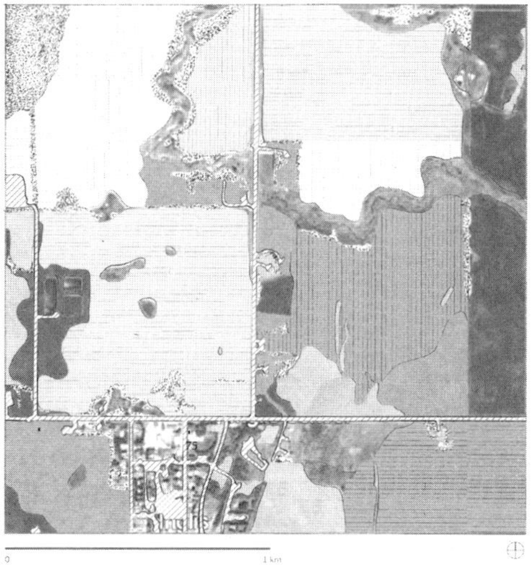

Andrea Gorda and north-side shelterbelt, 2020. (Photograph by Brenda J. Brown)

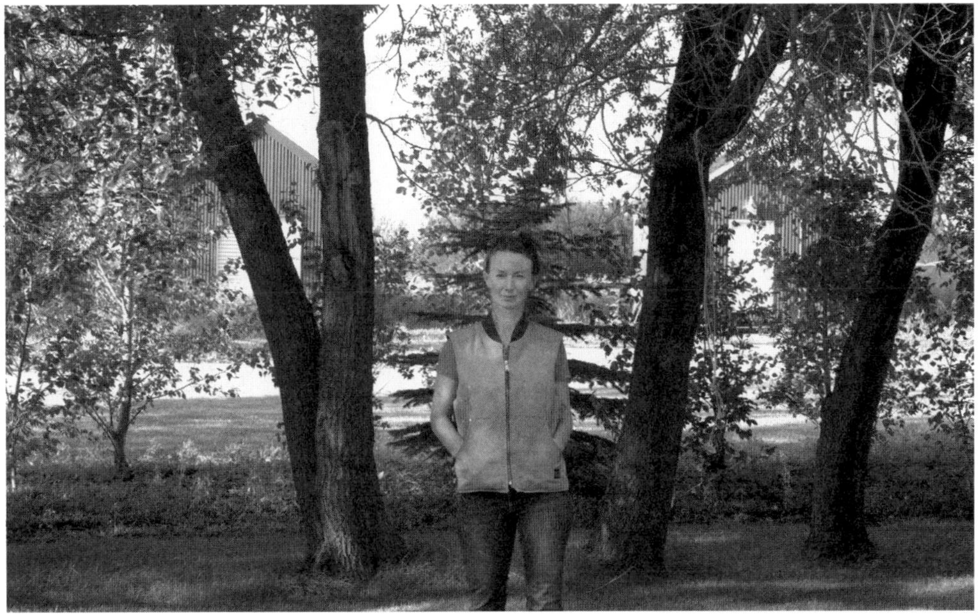

no remembrance there. It's just acres for dollars, you know, and so that's kind of the things that I think about when I'm driving around.

It was the yard that sold us because when we purchased the house, it was covered in Tyvek wrap. It was full of mice. And it was just horrific. Not that it's anything great now, but it's perfectly habitable. And if we didn't have the shelterbelt, we'd be like our neighbors who are constantly fixing their shingles. I mean, it depends on the aspect, on which way the house is facing, with how it's catching the wind, but they fix their shingles like three or four times a year because the spring winds that come through here are just horrific.

I think a lot about how the landscapes are constantly shifting. Especially when you continue north on Highway 83 and you go down into the valley and how it's all just, I don't know, but it's like we've had such a huge effect on the landscape but yet the landscape keeps moving, especially when the hill slumped, and the highway broke. It's like this tug of war in this movement back and forth between our culture and who we are as people and then the land and the plants constantly trying to reclaim . . .

AN—*Acer negundo* / CA—*Caragana arborescens* / PP—*Picea pungens* / Ph—Populous hybrid / UP—*Ulmus pumila* / SV—*Syringa villosa*

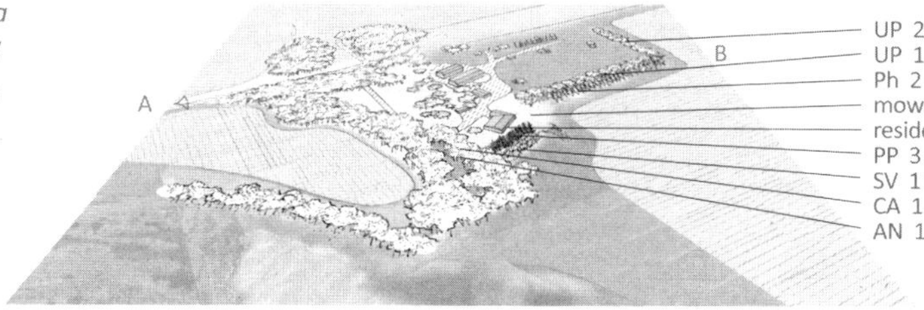

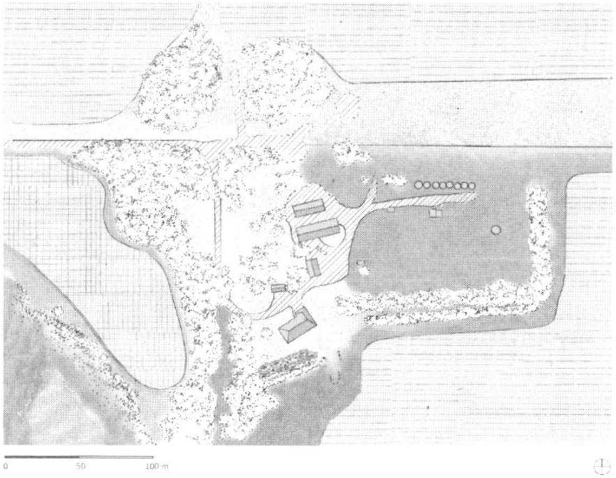

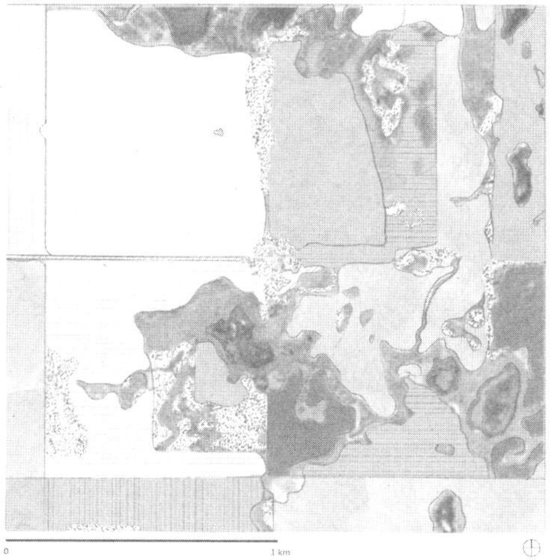

Entry to Muir farmstead.

A

Robert, Dolores, and Leanne Muir

Robert: We've been in this farmstead since 1976.

Dolores: Forty-four years.

Leanne: And I grew up here. I lived here my whole life until I went away to university when I was eighteen.

. . . How old were you when you bought the farm?

Robert: Twenty-three. We got married in May of '76. And I don't know if Dolores even saw where we were going to live at the time. (laughs)

Dolores: You know, when we saw this place it was a MACC property and we were looking for some land . . .

B

Entry to Muir farmstead.

Robert: MACC—Manitoba Ag Credit Corporation was an NDP [National Democratic Party] program that would buy the land, and they would rent it to you forever and ever and you'd never, ever get ahead . . .

(laughter) . . .

So, we bought the land and we said, oh, great. There's a house on it. And there's trees around here. We'll move here. This will be good. We bought the land, and after we bought the land we went into the house.

(all laugh)

Dolores: People would never do that now.

Leanne: Sort of bought it without really seeing it.

Dolores: And we were very lucky that it was well treed . . .

Leanne: *(laughing)* It happened to come with trees. . . . And it happened to come with a house.

(all laughing)

Robert: And there was no door on the porch.

(all laughing)

Robert: And there was a barn—full of manure.

Dolores: So we were very lucky. A lot of people couldn't get that lucky.

Robert: We were a happily married couple.

The trees didn't really figure into the equation at all. Basically, once we got here, we started looking, "well they are pretty nice." And people driving in, "oh what a beautiful place you have" and it was. It was really nice, you know. . . .

Dolores: But the yard used to be so sheltered with maple trees. You couldn't see down the lane and we actually named our farm Maple Lane Farms.

There were some evergreens here, too, that were fairly old, and some fell down all the way. We have old photographs of the farm, where there was just big rows of evergreens. . . . It was quite a while ago, but the people who used to live here came driving by one summer. I wasn't home but I think you told me.

Robert: Yes, many, many years ago.

Dolores: Yes. But they wanted to see the yard and they were saying, "Oh, we remember planting those trees." You know, I wish I would have been here to hear them. . . .

Robert: The farmstead shelterbelt is aesthetic, basically. It's nice to kind of look at it, especially if it's green, and we like to see the birds. And it attracts birds as you can see—there's a real pretty little redheaded bird there—and also it helps break the wind. You get northwest winds in the wintertime and it's not very nice if you're sitting out in the open. . . .

Leanne: The stand of Manitoba maple that used to be here, that's where I played the most when I was little. And there was this one maple that had fallen in the bush and I would always play underneath that maple because it was down and it was a huge tree, and it was a perfect size for kids—for me—to climb up onto it and play underneath it. And that just felt like it was so far away from the house even though it was right there. And it was my place. My playhouse was in that forest. I call it forest. Really it was for me.

Dolores Muir, Poppy, Alex Sawka, Robert Muir, Aiden Sawka, Leanne Muir. (Photograph by Brenda J. Brown)

I do remember though that with the rows of trees that used to be back . . . well they're still sort of there where the caraganas are—there used to be more of the maple trees and that's where the snow would drift in the wintertime.

Dolores: Oh, yes.

Robert: Big banks.

Leanne: And we'd get huge banks of snow in there. And we'd build snow forts every winter and it was just so much fun. . . . I think it created that nice place for all of the snow to accumulate that provided endless hours of winter play for us and we would hollow those snow drifts out and make . . . Mum, I remember you hollowing them out for us and digging in those snow drifts . . .

And we would spend a lot of time in the snow banks in the wintertime. And we don't get that amount of snow anymore. The trees have thinned out and the snow just doesn't accumulate in there in the same way.

Dolores: Yeah, but there're still nice snowdrifts in there.

Leanne: Yeah. But they were huge when I was small—so who knows how big they really were. . . .

Karen Klassen

I've lived here for almost, I guess, two and a half years. but I actually grew up here. So I lived here for sixteen years in my childhood. . . . Currently, it's technically myself, and then Dad's here during farming seasons and Mom will come sometimes.

My parents moved from here when I was sixteen and changed professions. So he's been a part-time farmer since 1993. And he would come back for harvest and seeding, but my uncles farmed for him and did a lot of the day-to-day work. . . . And then I went away, twenty-five years I was away. I went to university, and then I ended up moving to London, and I was a dietitian in London and then doing research. And then I got my PhD in Australia, and then Dad started talking about retiring from the farm, and I couldn't handle someone else farming our land. So I thought I should come back and learn how to farm. So now I'm a nutrition-scientist farmer.

My great-grandfather purchased this land for my grandpa and his brother in . . . between 1936 and '39—and they lived across the road. And then when my Oma and Opa got married in 1941, they built a house just around the corner. . . . That's when they started planting trees and building this as a whole, I guess, yard. . . . And then in 1977 this house was built. And then my parents lived here and they started landscaping the rest of the yard.

We farm grain. We used to have hogs in the '80s and '90s, probably up until mid-2000s,

CA—*Caragana arborescens* /
FP—*Fraxinus pennsylvanica* /
Ph—Poplar hybrid / PP—*Picea pungens* /
SA—*Salix alba* or *Salix* acuifolia /
UP—*Ulmus pumila*

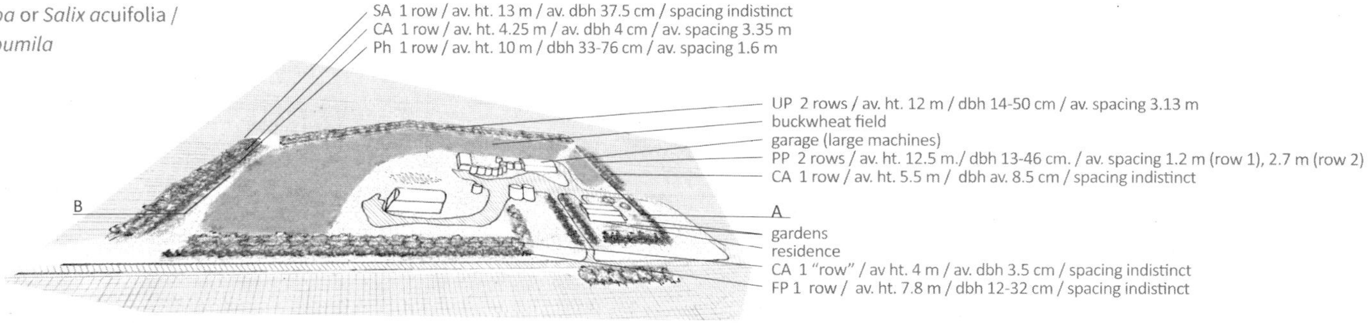

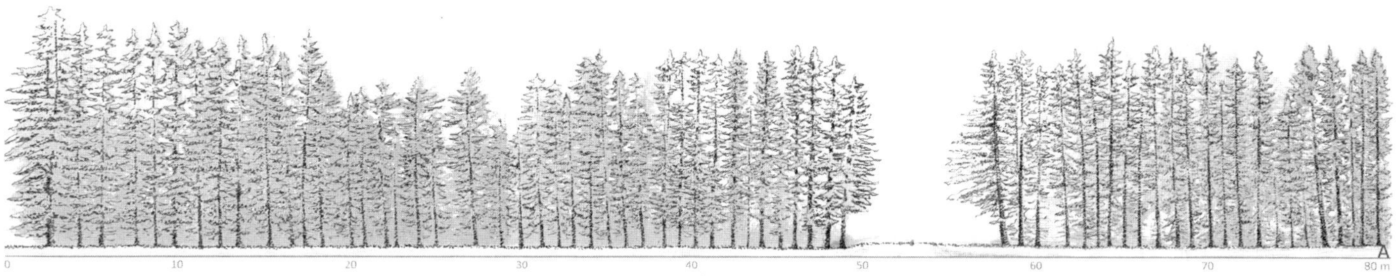

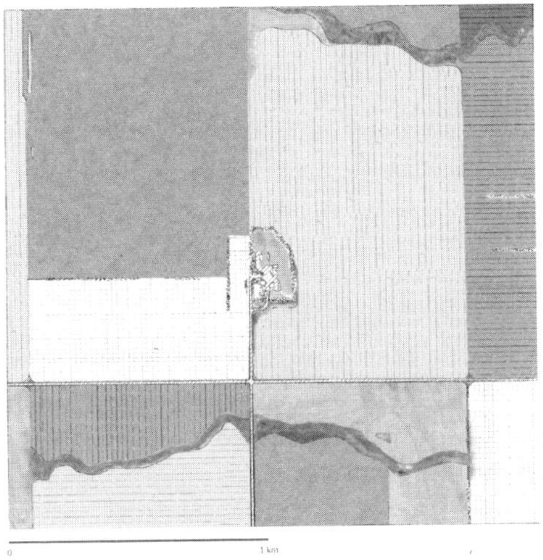

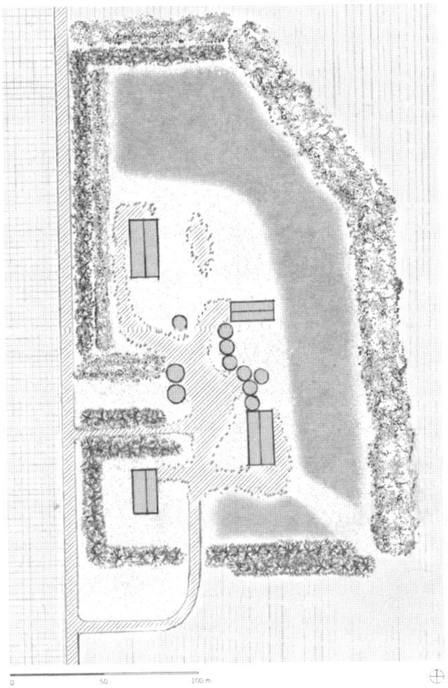

but now it's just grains. This year we have wheat, oats, barley, rye, canola, peas. . . . We've got some land going into organic agriculture so we've got a green manure. . . .

The first thing for why I'm enthusiastic about the trees and why I'm planting more is just the ecological goods and services. I'm very well aware that we are stripping our landscape in the way we are currently farming; it's not an ecologically beneficial practice and we need to do something to mitigate. So that's the first aspect. But from a protection perspective, to me personally, it definitely is for the house but also all of our buildings. We just put up new bins and I just saw someone with no shelterbelts that had all their bins fall over from the wind. Our winds are so insane here that you absolutely need some sort of buffer to protect that as well. Last year, we had that big snowstorm in October, and the day before I had a deer in the yard. We have deer all the time but they always come at night and last year they came during that day and I don't know if they sensed something but they were sitting in our garden. The next day during the storm I had six deer wandering around during the storm you know, benefiting from our shelter in the yard. And it was just beautiful and it made me realize all the additional benefits these protective services can offer.

We've willows on the north side and the rushing of the leaves and the sound, you know, the ocean sounds in there is beautiful and it's very calming and peaceful especially if there's a light wind. It's just stunning to be around. . . . The fact that it's providing housing for birds and tons of other animals is fantastic as well, just to be around. I always see butterflies. I even see Monarchs . . . I always see them hanging around the trees. I've got a ton of pollinator flowers all throughout my gardens, and I definitely see them there as well. But I always notice them around the trees. . . .

I'm doing lots of herbal teas and so I'm actually harvesting spruce needles as soon as they bud for teas and other sort of things, and I'm harvesting willows for herbal medicines and stuff which definitely was not the intention when they were planted.

Those two rows of trees, in the summer it's hard to get through. But as soon as the grasses die, it's this beautiful path. . . . this magical little alley-way. . . .

I think the shelterbelts in the landscape make part of our identity. . . . You can tell when you're in Manitoba, because everyone has a very similar sort of style, you know, different species of trees, but it's a similar pattern of landscaping. And it's odd. When you drive down the highway and you see something without a shelterbelt, it sort of sticks out as a strange

Karen Klassen and west-side shelterbelt.
(Photograph by Brenda J. Brown)

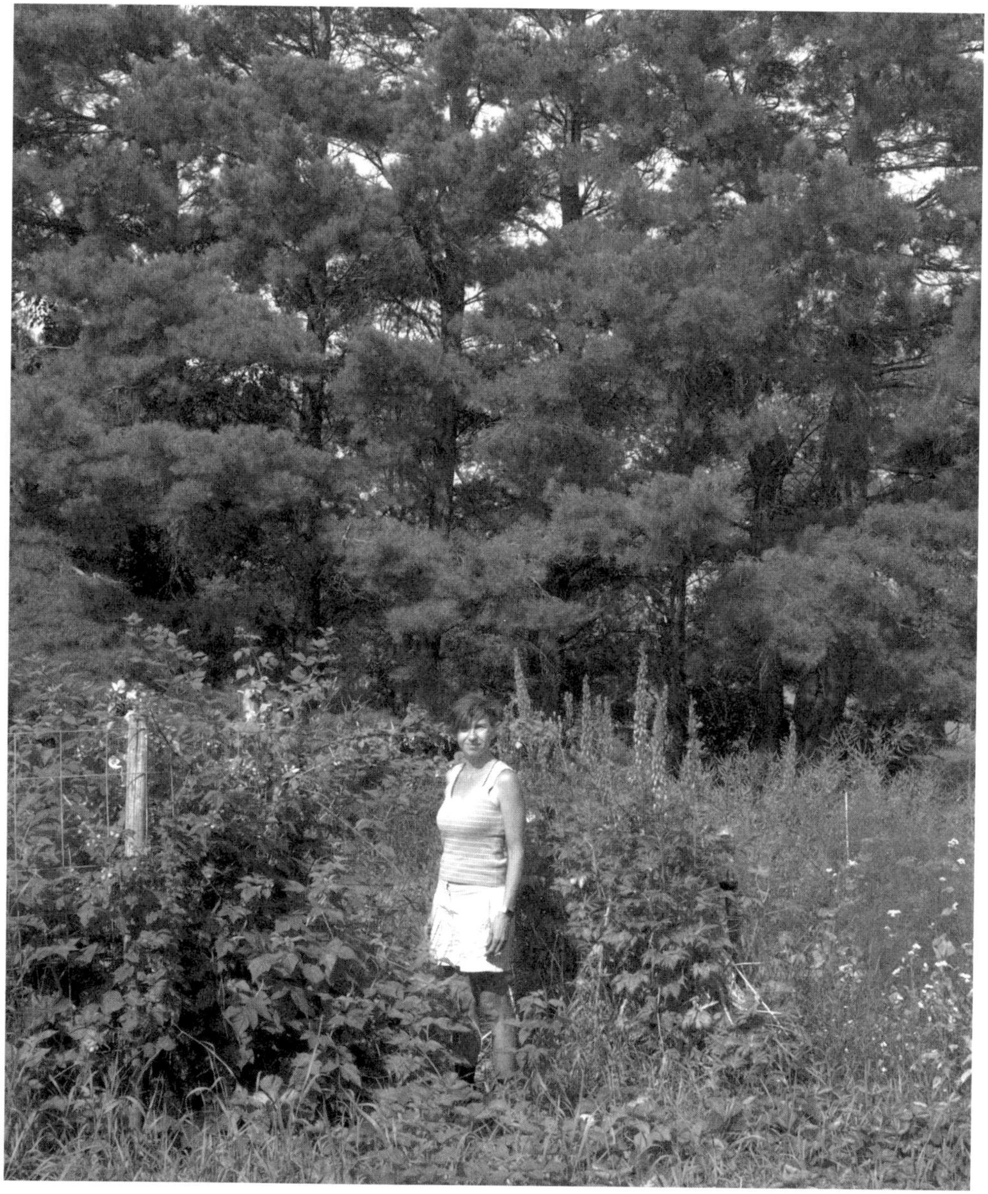

Klassen shelterbelts in winter. (Photographs by Karen Klassen.)

phenomenon. So I think it seems like at this point, since Europeans populated this region, they've been part of the culture of creating your own little yard site.

I've seen a lot more of the field shelterbelts go, even last year, like my neighbors just tore all theirs down, but they're keeping the yard. So there is definitely still that connection and values seen to it.

My oma and opa had their house over there and they had their old garden there and the only things left are the two crabapple trees planted in the '40s. And I definitely feel a connection to that, probably more so than to the spruce that he planted very strangely and way too close together. And of course if you plant spruce close together they may provide shelter quickly but if you don't thin them out they get ugly and sad on the bottom. I think I totally understand the family connection because there's stuff around that you always think about, and it's kind of funny. It's good, you know, remembering the funny things.

CA—*Caragana arborescens* / FP—*Fraxinus pennsylvanica* / PP—*Picea pungens* / SA—*Salix alba* or *Salix acuifolia*

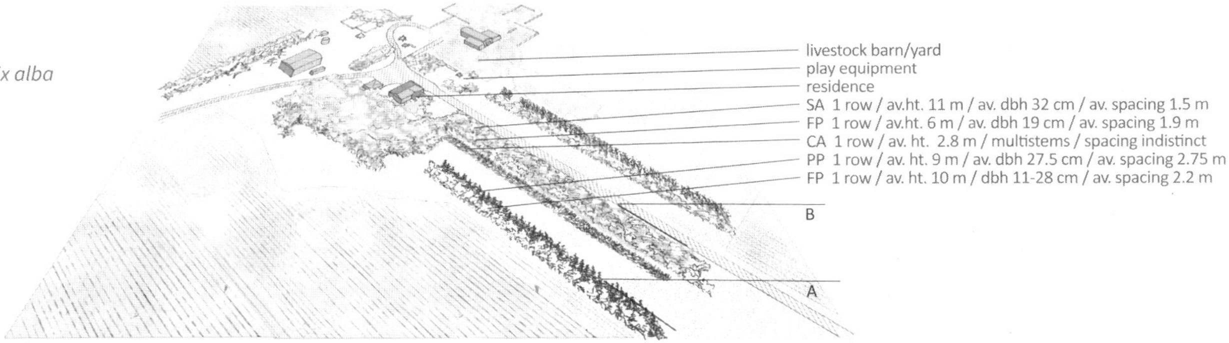

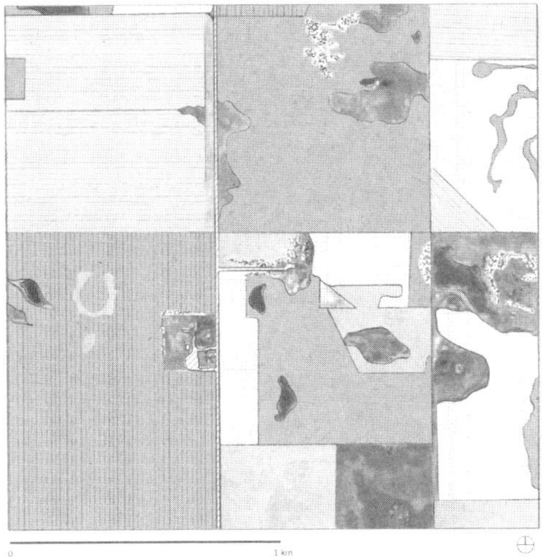

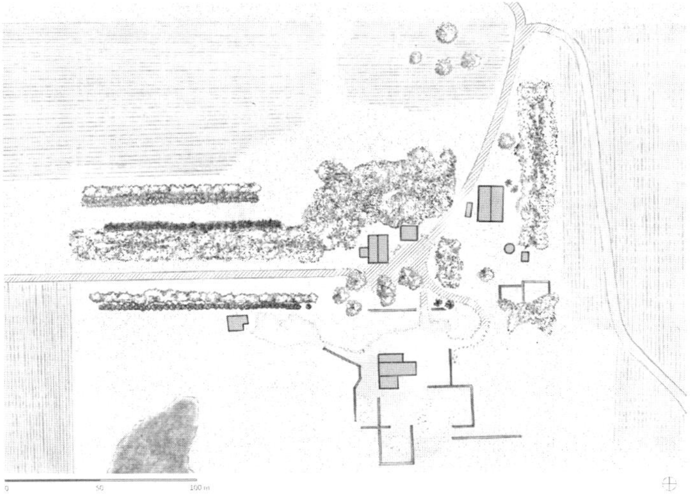

Scott Hainsworth

I moved here when I was five, so that would be twenty-five years ago.

I don't live here; I live in town but this is the farm I grew up on. So this would be my dad's farm. . . . This is kind of just home base, to raise some cattle and grain and that kind of stuff. Maybe 50 percent, 60 percent of the time there's . . . somebody here every day but everybody's in town—Hartney—now. But I grew up here; I went to school from here and everything. . . .

Originally there was another row of trees here. They were kind of aged out or dying. So they were pushed with a cat and a backhoe and burnt and buried. All along here there were large maples and ash trees. . . . And then I guess when we were old enough to work a hoe and water and weed, Dad started planting trees and that's what you see now. It took them about three tries, I think, you know, one year it was just too dry and they died. Yeah, I think it was two years of kind of drought-like conditions till we finally got a catch of them, and we got what we have here now. So it took time . . . it doesn't happen overnight.

. . . My brothers and my sister and I, we'd have to get up first thing in the morning, like six o'clock, before they got the heat of the day and we'd hoe them and weed them, and then in the evenings we'd haul water and we'd walk behind the grain truck with a hose and a water tank and water them and that was our maintenance to keep them alive. . . .

I be the oldest and then I have a sister; she's two years younger than me, and then a brother that's five years younger, and then a brother that's seven years younger. . . . So it was a great bonding time I guess you could say. It wasn't what you wanted to do, but I guess you could say maybe you're thankful for it later that it taught you a lot of work ethic and how to get stuff done. So I guess, as we're older, we're maybe grateful for it, but at the time not so much.

We, the three of us brothers, still farm on the side. Farming doesn't really pay the bills unless you're a huge farm. . . .

Scott Hainsworth and north-side shelterbelts. (Photograph by Brenda J. Brown)

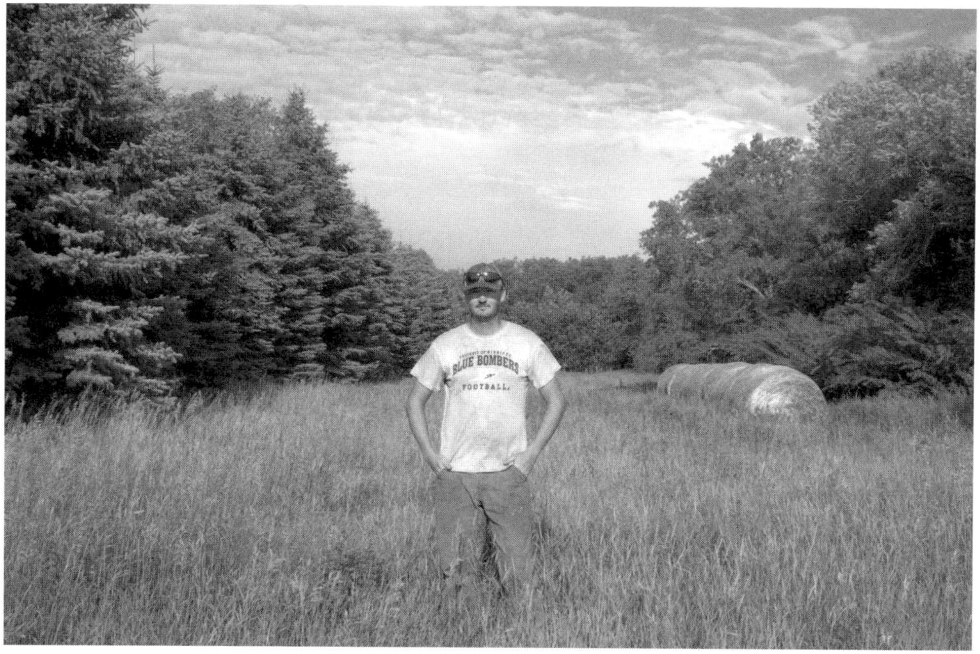

We planted some trees this year on the west. It's really open to the west and our goal is to maybe, you know, slow some winds down 'cause in the dead of winter when you're calving cows, you kind of want to have as much heat source around your barn or your calving unit. So if we could slow some wind down . . . but it they're probably fifteen years away from doing anything; they just take time.

A cool thing is, you know, when the Homesteaders Act came through there was probably a farm on almost every quarter, right? And if you look around now the only yard sites that are left seem to be yards that have really established shelterbelts on them. . . .

Most of the trees were planted in my lifetime. The willows along the road there are older—they're so thick that it's hard to tell where their trunks are coming from—what trunk is going to what tree. And there's poplar. These are green ash. There be lilac. There's maple. Those ones with the cones, those are caragana. There's some ponderosa pine. And then the newer ones are ash and blue spruce. And the ones we planted this year are a hybrid willow, blue spruce, and then there's some velosa lilac as well. And then if you look further to the north there—those maples just seem to grow like weeds around the yard. And same with the cottonwood trees. They just kind of grow where they want to grow.

When you plant trees you're tied to, you know, all that effort that you put into starting it from a little seedling . . . you know a lot of trees outlive people. Say, this is "my Dad started it" or "it could have been grandpa's before" and so there's a little bit of heritage like "this is the family's yard" and I think people see some value in that. Maybe that's just my spin on it but, there's maybe a lot of work that went into it, or Mom put a lot of time into mowing around them, or, if you knocked them all down your sister would kill you or something like that . . .

I just remember this seemed like a mile when you were a little kid eh? This seemed like a really big shelterbelt, but really it's not that big.

Bernie van Kemenade

I've lived here since 1967, so fifty-three years. In 1860, which was the year of the big flood [from] where Grosse Isle got its name—because it was an island—this piece of property, this section, was the first deeded section, and actually it was owned at one time by my wife's great-great-grandfather. And then he lost it during the Depression.

My father bought it in 1967; I was fourteen years of age. And then I bought it from my father in 1983. But I've always lived here except for a couple years when we first got married. I'm retired now but we used to farm—originally a dairy farm where we grew lots of hay and feed grains, and then it developed into a larger grain farm. And I retired when I was sixty-one and I rent the land out now. We were growing everything from canola, corn, sunflowers, soy beans, [to] wheat, barley, oats, flax. So just generally, cereals and oilseeds.

My brother-in-law rents my farm now. . . . I used to farm with him and then we split when I wanted to retire, and he kept going. So he rents my farm now.

[I raised my family here, but] it's just me and my wife live here now.

Most of my property is natural bush. We did have a lot of trees planted that my dad planted in the '70s. You can see some of them there. I took two out of three out, I left every third tree. They were planted too close and that's a problem most people make or a mistake most people make when they plant shelterbelts. When the trees are a foot high, they look very far apart, but they're actually way too close. And you end up having to take a lot of them out.

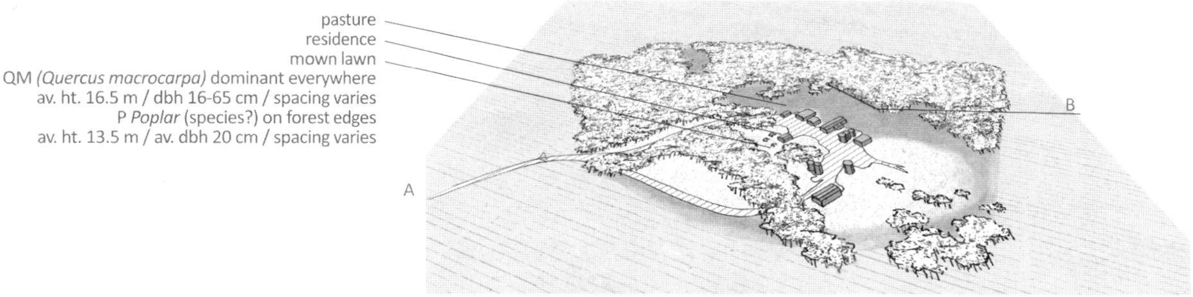
pasture
residence
mown lawn
QM *(Quercus macrocarpa)* dominant everywhere
av. ht. 16.5 m / dbh 16-65 cm / spacing varies
P *Poplar* (species?) on forest edges
av. ht. 13.5 m / av. dbh 20 cm / spacing varies
A
B

0
1 km

0
100
200 m

B
0
10
20
30
40
50
60
70
80 m

. . . We broke about one hundred acres of bush when we first moved here and turned that into farmland, but we have sixty acres in our area here. There's a few cattle in here now—they're not mine, I just let him use it. Yeah, we like the natural bush. Natural bush is a lot tougher species than anything you plant we found out, like when it comes to farm chemicals you have to stay so far away, but natural bush, it doesn't seem to bother them. . . .

These oaks, they're great trees . . .

I'll bet you this first tree here is five hundred years old. Yeah, black oaks live to be five hundred . . . they're an amazing tree. I've watched oak trees all my life. I mean, if you could plant shelterbelts out of them, they'd be the best tree. Not much hurts them. The wind doesn't bother them much. They live forever. You know, you get lots of leaves off of them, but we can deal with that. But they're just a great tree, they heal themselves. They'll stand for five years without a leaf on them. And then all of a sudden they'll start growing again. While they're healing themselves I don't let anybody cut anything down unless all the bark has fallen off; then I know that tree's dead. 'Cause I can show you trees that have come back after five years; I can show you how they heal themselves and they're an amazing tree.

Well, I let people know when they eventually fall over, you know, we take them for firewood. I don't burn wood myself, other than for the odd campfire, but I do have a couple people that come in and cut down dead fall and that just to keeps it out of the road.

. . . You know if we had to do it all over again, we would have left a little more bush in the front here. But you know where you came in on the driveway is where it should have been cut off, that would have given us a little bit more bush. But yeah, I think it puts you a

A

Bernie van Kemenade and oak tree. (Photograph by Michaela Peyson)

little more in tune with nature when you live in . . . it's like city people going camping in a national park or provincial park. You know, but when you live in it every day, you see what goes on in the bush, the bush is alive all the time. . . . And I like that. That's why we've never bought a cottage. Why would we go live in the bush somewhere when we haven't got a neighbor for a mile around? So you know, we like it here. As long as we can stay, we'll stay here.

Lane van Kemenade

I've lived here for five years now.

This was my grandfather's house before me. He built it in 1981 and planted every tree on the whole property. My wife and I and our little dog Scout live here. Just the two of us, three of us, I guess.

I do not farm. I used to. But now I've started a construction company and I run the construction company out of the back of my dad's farm yard . . . heavy construction with excavators, dozers, and gravel trucks, mostly ag construction for the local farmers. . . . We work on barn pads, shop pads, driveways, road building, ditch digging, land clearing, site prep, and excavations and trucking.

The house was built in 1981. And I believe all of the trees were planted the year before in 1980, so that's thirty, forty years old now. My wife and I have replanted about a dozen different trees in the last five years, just trying to keep up . . . sometimes a lot of them are dead and dying. Now they're getting towards the end of their lifespan, some of them, and so when they die, we like to replant, and keep growing new trees, new stages.

We trimmed down one row recently and it seems to let more of the wind in now. We noticed a bigger snowbank this spring, just from catching the snow. Our greatest protection has been the caragana, so the outside row. They really do catch a lot of the snow before it comes in here. And sometimes when we check the weather, inside of our yard, inside of our shelter belt, it seems nice and warm and calm. . . . And sometimes it's tough to judge the weather when you're going out, what you want to wear. You really have to go out to the end of the road to kind of get the true feeling for it sometimes, but yeah, it's quite nice. We quite enjoy it here.

I've come here all my life. My grandparents were here and we'd come across the field on a little trail. And we'd come here for milk and cookies and all the treats and I remember when the trees were just my height, maybe six, seven, eight feet tall. And now some of them are thirty, forty, fifty feet tall. And I've watched them grow all my life, and they were planted six years before I was born. But, in my teenage years, I always remember coming here and thinking, oh, wow, what a yard this is. We get quite a few compliments on the yard.

. . . All the evergreens are double spaced, except for the front row. And now we've come into problems years later, where some of them are too close together. And we're almost at the point where we have to sacrifice every second one in order to expand every other one. I don't

CA—*Caragana arborescens* /
FP—*Fraxinus pennsylvanica* /
PP—*Picea pungens*

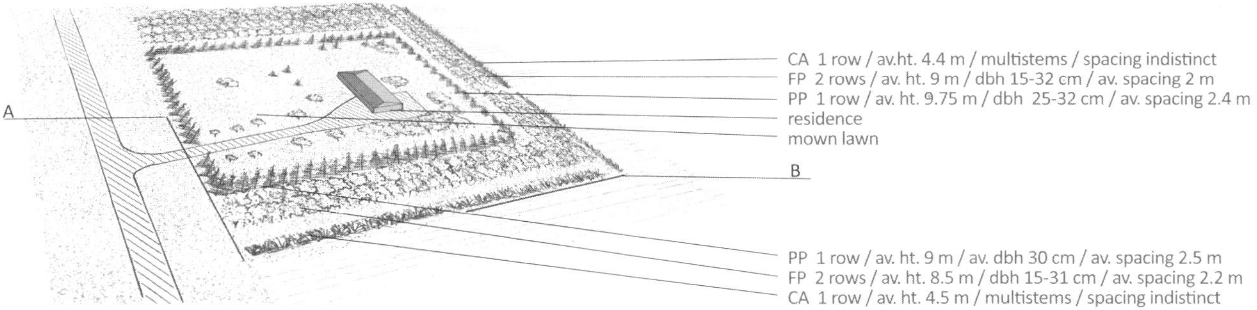

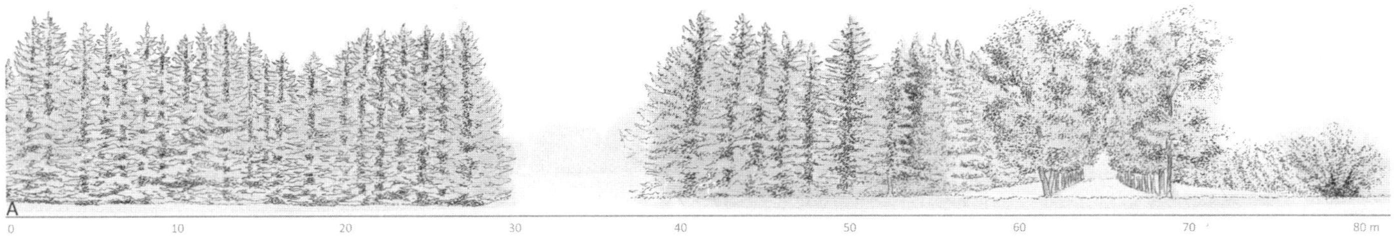

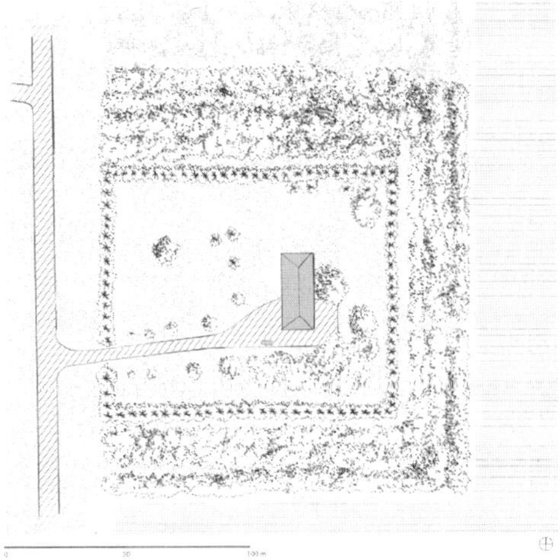

think [my grandfather] intended this to get as big as and beautiful as it did. He was just built out on the open field and wanted some shelter. And the best way he could come up with was four nice straight rows. If you look from the air, every row in this yard is straight. . . . With my drone I fly up and I get an overhead view from five, six hundred feet up. And it's very neat. The way he planned it back in the early '80s with no technology, just a string and a couple posts.

. . . There's a lot of elements out here in the bald prairie. . . . We're quite lucky to have such an established shelter belt already. And it makes me want to establish more shelterbelts, or more trees, more protections for the future generations as well too. Because I was lucky to come into this from my grandfather. And I want to be able to do it one day, to say my kids or my grandkids. I was involved in planting the shelterbelt at the carwash when I was . . . ten or twelve years old, maybe. I remember being out there planting these little sticklings and once you do that, you kind of have a connection to them. . . . It's kind of humbling when you talk about it like that, actually. . . .

We just put up a bunch of birdhouses. I'll show you there later if you want to go. And it's very neat just to walk through and to drive through and take in the nature part of it. And yeah, we quite enjoy it.

NOTES

This work has been generously supported by the UM/SSHRC Explore Grants Program and a Canada SSHRC Insight Development Grant. In addition to the participants cited in this article, I am grateful to Ute Holweger, Henry de Gooijer, Blair English, Jason Hare, Dean Brooker, Jeff Theil, Cliff Greenfield, Melissa Sturgeon, Zachary Sturgeon, Jeff Kostuik, Don, and Jan for their help.

1. The obviousness of this progression does not escape me.

2. Update: In summer 2021 eleven more Manitoba farmsteads were documented and their owners interviewed, one of them a Hutterite colony minister. This means a more extensive and thorough representation of southern rural Manitoba—including properties in the Inter-Mountain, Whitemud, and Redboine watershed districts—will found in the final project works.

3. Properties from this time were laid out on the French river lot, or long lot, system, a pattern still evident in Winnipeg's street network today.

4. This can make for some navigational complications today, as Canada roads and mileages are now measured in metric units.

5. John C. Lehr, "Settlement: The Making of a Landscape," in *The Geography of Manitoba: Its Land and Its People*, ed. John Welsted, John Everitt, and Christhoph Stadel (Winnipeg: University of Manitoba Press, 1996), 92–101; James M. Ricktik, "The Township and Range Survey System," in *The Geography of Manitoba: Its Land and*

Carly, Scout, and Lane van Kemenade in shelterbelt allée. (Photograph by Brenda J. Brown)

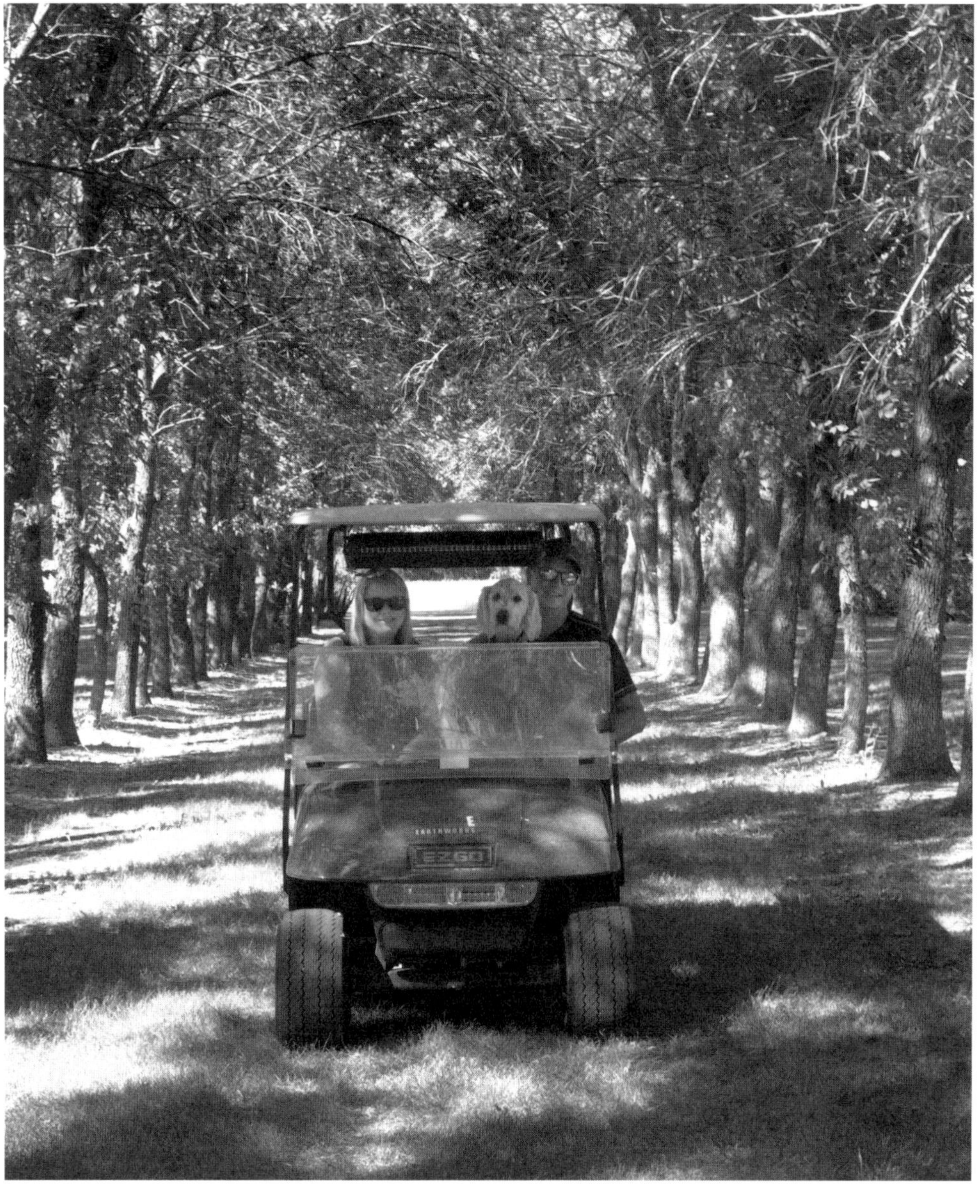

Lane and Carly van Kemenade's shelterbelt canopy in wind (sequential images). (Photographs by Brenda J. Brown)

Its People, ed. John Welsted, John Everitt, and Christhoph Stadel (Winnipeg: University of Manitoba Press, 1996), 102–3. This settlement campaign had significant, unequitable implications for different groups, most obviously for First Nations and Metis people who had called the lands of Manitoba home for generations. However, serious discussion of this history is beyond this essay's scope.

6. Various writers have commented on the tendency, as Clare Cooper Marcus put it, to reproduce "special places of childhood in the adult home." See her "Environmental Memories," in *Place Attachment*, ed. Irwin Altman and Setha Low (New York: Plenum Press, 1992), 98.

7. N. V. Thevathasan, B. Colwman, L. Zabek, T. Ward, and A. M. Gordon, "Agroforestry in Canada and Its Role in Farming Systems," in *Temperate Agroforestry Systems*, 2nd ed., ed. A. M. Gordon, S. M. Newman, and B. R. W. Coleman (New York: CAB International, 2018); Suren Kulshreshtha and Edward Knopf, "Benefits from Agriculture and Agri-food—Canada's Shelterbelt Program: Economic Valuation of Public and Private Goods," Report (Indian Head, Saskatchewan: AAFC Shelterbelt Centre, 2003).

8. Thevathasan et al., "Agroforestry in Canada and Its Role in Farming Systems."

9. The ash trees that were long part of that shelterbelt palette are now severely threatened by the emerald ash borer.

10. Kulshreshtha and Knopf, "Benefits from Agriculture and Agri-food—Canada's Shelterbelt Program."

11. William Ashton and Gillian Richards, *2014 Shelterbelt Survey: Prairie Producers Use of and Attitudes toward Shelterbelts* (Brandon, Manitoba: Rural Development Institute, Brandon University, 2014).

Monk's Garden, Isabella Stewart Gardner Museum, Boston. Designed by Michael Van Valkenburgh Associates. (Photograph courtesy of Michael Van Valkenburgh)

IV. Gardens, Technology, Fantasy, Experience

Hidcote Manor Garden. (Photograph by Robert B. Riley, courtesy Bob Riley Landscape Architecture Collection, Digital Collections at the University of Illinois Urbana-Champaign Library)

1988

From Sacred Grove to Disney World

The Search for Garden Meaning

ROBERT B. RILEY

A garden is about nature. The meaning of nature itself is a large issue, particularly for our time. Most of us assume that nature is inherently good for people and that it carries a universal meaning. Garden is seen as a special expression of that meaning. It has been an expression of importance in most cultures and the highest art form in some. It is an expression that, after several decades of being ignored by designers, has again become a focus of work, debate, and fashion.

Garden is a fuzzy or pluralistic concept. It includes three essential features: a thing grown (for pleasure or profit), the activity of growing (gardening), and a place (the garden). These total to a garden, be it a patch of tended vegetables or Vaux-le-Vicomte. What makes a garden a Garden, that carrier of meaning and high art form that concerns designers, is the intent to create an ideal environment, a place of special qualities, of itself or by allusion. Understood in this way, the garden is an intense and particular statement about our relations to nature, an archetypal statement of those relations.

It is not the only such statement. The idea of garden becomes clearer in contrast to other archetypal relations with nature. Hunting is one such relation, an activity of long tradition, glorified in literature. Paul Shepard (1959, 1973) and José Ortega y Gasset (1972) have written eloquently on hunting as the prime relation with nature, the one that makes a man truly a man.[1] On a less abstract plane, hunting as love of nature is a major theme in the works of Aldo Leopold (1964). No matter what one might think of it morally, hunting is a relationship that, at its purest, demands attention, involvement, dedication, knowledge, skill, and discipline in relation to nature.

Animism, too, is a distinct relation with nature, whether in the sense of assuming a soul within every organism or in the sense of that intense, transporting response to nature, distinct from the child's wonder or the adult's analysis, as described by the author and amateur naturalist W. H. Hudson (1942). Animism is a relationship not antithetical to the garden, but surely different, an emotion that can be experienced in the wild, or in a garden, or in any natural setting.

The wilderness experience, today a mystique, is another such relation. Maybe this immersion in "untouched nature," eloquently stated in the writings of W. O. Douglas (1961) and numerous Sierra Club publications, is contemporary animism with moralistic overtones. Maybe wilderness is a setting polar to the garden; maybe, as J. B. Jackson (1980) has stated, it is not. It is certainly, however, a relationship with nature that has both similarities and differences with the garden, and a relationship that has a large, articulate following in American society. As such it has directly affected our attitude towards the garden.

Pastoralism shares with garden the vision of an ideal, harmonious relation with nature, in a setting affected by human presence. The difference is that in pastoralism human activity is but one element in an extra-human or natural order. In a garden human intervention is the prime ordering element. As Hugh Johnson (1979, p. 8) has told us, "The essence is [human] control."[2] Yeats's poetic vision of the pastoral, "The Lake Isle of lnnisfree," contains a garden, but it is only one element reinforcing a larger gestalt.

This essay is a simple and naïve pursuit of meaning in the garden. It is appropriate, then, to use the simplest, most naive and inclusive definition of the garden that can distinguish the garden from other relations with nature. For our purposes, a garden is an attempt to *establish meaning by giving form to nature.* This definition, if not better than several others that might be offered, fits the consensus about that elusive core of meaning and is particularly appropriate given the title of this volume.[3] It adequately sets garden apart from the other archetypal relations and settings mentioned. It does not, however, distinguish the garden from the park or from some forms of environmental art. This is a non-trivial issue that will be discussed.

Themes in Garden Meaning

In our search for garden meaning the first question to ask is not what a garden should mean but what gardens *can* mean, what gardens *have* meant. A large catalog is at hand for answers, for the volume of literature on the pleasures of the garden probably equals those devoted to the parallel pleasures of the table and the bed, including illustrated, step-by-step instructions. (The parallels among garden design, cuisine, and sex—including the roles of creators, connoisseurs, and consumers—is a subject worth exploring, but not here.) The most complete and literate compendium of garden uses and delights is contained in the first and last chapters of Nan Fairbrother's *Men and Gardens* (1956). Walking and talking and growing things; renewing jaded senses with multiple, ever-changing stimuli; providing seclusion for eating and playing games and making love; humanizing our geometric abstractions with the "background of green life";[4] giving to children the spur to fantasy and make-believe and to adults a glimpse of the sacred, the immortal, and the heavenly; and providing safe, harmless occupation for prisoners and women, a therapeutic pastime for invalids and the retired, and nostalgia for all. The garden, Fairbrother tells us, is a supportive and adaptable setting for many of our activities and our aspirations. In a practical approach to the difficult issue of meaning, she notes that gardening has always been thought of as exceptionally virtuous but adds: "I cannot see why growing flowers is a more virtuous way of spending time than taking opium, or any other selfish way of seeking happiness. But I can very well see that it is a more lasting pleasure, more socially acceptable, and certainly more healthy" (Fairbrother 1956, p. 8).

More than useful, more than delightful, the garden is often seen as a primal human setting. Jackson (1980) has included a garden with home, road, and shrine as one of the four landscape archetypes. Olin (1986) says that garden is to landscape architecture as home is to architecture. The comparison of home with garden is illuminating: both have origins combining the sacred and the domestic, garden is often seen as mediator or transition between home and hostile outer world, and both are fuzzy or pluralistic concepts, persistent but variable through time and across cultures.

The essence of garden, that core responsible for its persistence, is most commonly thought to be a spiritual-intellectual concept, an always mental, sometimes physical, artifact expressing humans' relation not only to nature but to their gods and their universe. This is what Genesis tells us. More recent writers such as Harbison (1977), Treib (1979), and Tuan (1974) see gardens as miniature recreations of the cosmos, models of an ideal world made from the stuff of the real. From this comes the garden's importance and power. But others, equally convinced of the garden's permanence and import, see in its core, primary statements about a people's relation to work, family, and society. Tuan (1984) also tells us that some gardens can be understood as the ultimate expression of centralized power and vain caprice. Jackson (1980) has considered the origin of garden as the family's private work space and told us that its opposite is not the wilderness but the communal field. Francis (1985) tells us that in contemporary communal gardens the important attractions are social interaction and work experience, and contrasts these gardens to parks. Now, when architects, always adept at justifying their designs by snatching intellectual phrases (e.g., the sillinesses perpetrated in the name of Eliade and sacred space), are joining the rush back to the garden, the unpretentious speculations of Jackson and the sensible studies of Francis should be recalled for our attention.

But whether its basis is social, spiritual, or psychological, the importance and persistence of the garden are not hard to understand. The garden is where we deal with the basic dialectic of our existence, the tension between life and death. Gardens, we all know, are about life and ever-promised renewal. Gardens, as Catherine Howett (1987) and Vince Healy (1987) remind us, are also about dying. Americans are good at ignoring death but earlier garden ages would have needed no such reminders. Robert Williams (1987) points out that a favorite theme of Georgian gardens was not just Arcadia but death in Arcadia. Sitwell (1909) saw in gardens the ghosts of dead civilizations and told us it is death to fall asleep there. Siegfried Sassoon (1949), musing on those same Italian gardens, also saw the ghosts of his poet predecessors, seeing those same ancient ghosts.

When we build a garden, we exert control over those forces of birth, growth, decay, and death that in the end are beyond our control. The impermanence of plants is not only the price, but the validation, of what we do. It reminds us that while our control passes, it is still control: control over not just clay or brick or steel, but over the stuff of life itself. This is why the hedges of a garden like *Chateau de Boleil* are more than green walls, and why a vine-covered pergola is not just a fragrant I-beam. To understand this essence of garden as manifest intervention in the processes of life and death, one might turn not to Versailles, as most writers suggest, but to Bonsai, as Tuan (1980) has recommended, in which the inevitable rhythms of nature have been slowed almost to a standstill.[5]

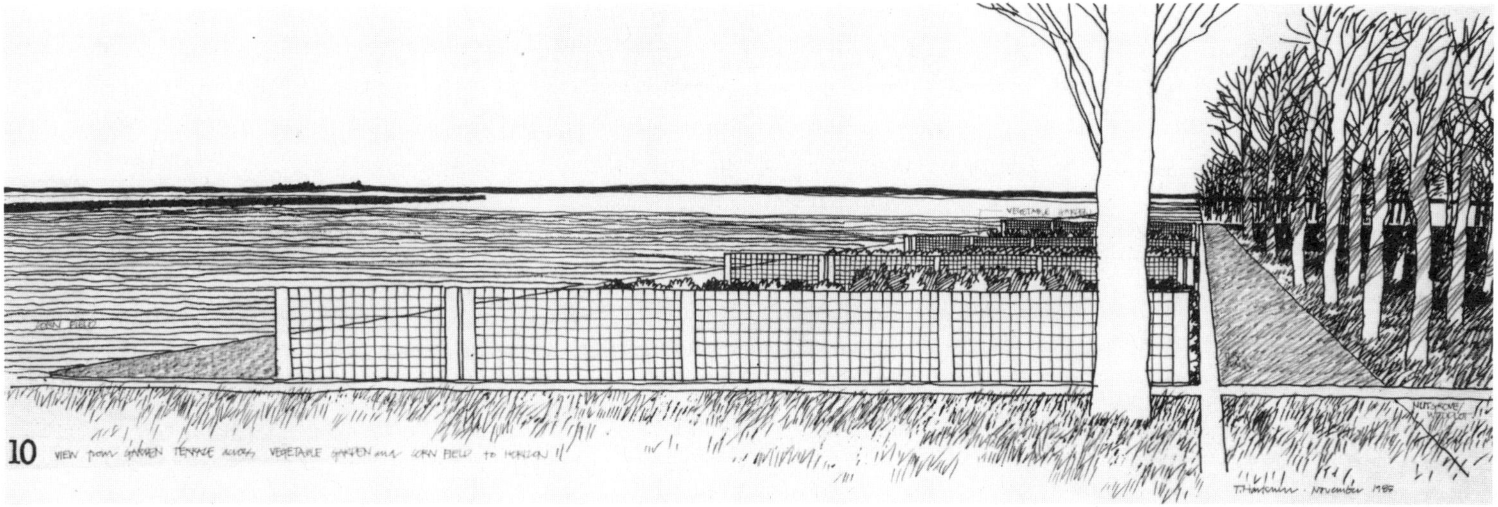

10
VIEW from GARDEN TERRACE across VEGETABLE GARDEN and CORN FIELD to HORIZON

Abstracting the larger landscape: a garden design by Terence Harkness. (Drawing courtesy of Terence Harkness, from original publication)

The garden not only expresses this, our most basic, dialectic, but layers upon it other dialectics: freedom/restraint, predictability/uncertainty, dominance/submission, nurture/neglect, order/disorder, variety/sameness, stability/change, and so on, even onto the tricky business of garden as itself/garden as something else. Kenneth Helphand's (1985) observation that pruning is the counterpart of planting, the wait of the gardener who can choose what to plant but not what will survive, the loose drift of Jekyll's plantings through the geometric order of Luyten's walls and terraces—almost all intense garden experiences can be phrased as dialectics.

Fairbrother (1956) has distinguished the garden from other art forms as the only one in which the basic elements are of intrinsic and complex interest in themselves. A plant is different from a brick or a tube of paint or a note on an oboe in three ways. It is a multisensory stimulus, it has a richly evident physical and visual structure exhibiting variety within order, and it contains many levels of information—from those of interest to the cell biologist to those simply wanting the cachet of a blue rose.[6] Wood and stone can be of similar intrinsic interest and structured complexity. Still, plants are commonly the foci of activity and interest in a garden—the reason for a garden's existence—while few but the obsessed architect or craftsman would consider a building's purpose to be the display of wood. However it might be explained, plants seem to be of inherent interest to humans.

But as designers we also know that a garden can be more than a collection of interesting natural elements; we know too that it can be a compelling spatial composition. As designers we also yearn for the return of those times when such compositions were even more than grand compositions, times when they carried meanings central to their society. These three roles of the garden have been described by Jackson (1980) as sources of delight, work of art, and symbol, and by Sun Xiaoxiang (1987) as nature's garden, the artist's garden, and the poet's garden. It is this possibility of being appreciated on three levels, combined with the availability of so many potential dialectics, that makes the garden such a common collector of individual and collective meaning, and such a potent locus for symbols that exist across cultures and through history.

The American garden has become such a symbol of genteel respectability and banal cheer. ("I count only the sunny hours") that we forget how frequently it has carried very different meanings. I have referred (Riley 1987) elsewhere to the garden as a symbol of power, with its corollaries of order, ownership, class, and status, and as a symbol of sexuality, with its associated themes of love, sex, and sexism. Readers interested in an example of the dominance and interactions of these themes might read Carol Fabricant's (1979) study of the philosophical underpinnings of the Georgian landscape, "Binding and Dressing Nature's Loose Tresses: The Ideology of Augustan Landscape Design." In Fabricant's analysis, the Georgian literary convention linking women and nature was also a statement of then current realities concerning politics and property. Both woman and landscape were badges of male power and possession; both were to be artistically disciplined for public display but wantonly permissive for their owner's private delight.

Few designers today would admit to shaping their gardens as expressions of political and sexual dominance. Still, it is this general role of garden as powerful symbol complex that today's designers seem so eager to restore. However diverse their approach, their common slogan is Robert Harbison's (1977) aphorism, "Gardens always mean something else."[7] Gardens have been a locus of meaning in many cultures, but not in modern America. Our "something else" got mislaid somewhere along the way. Before we join the rush to restore it, it might be helpful to inquire as to where it went, and why.

The Decline of the Garden

The decline of the garden as an important place carrying special meaning is not that hard to understand. A complex of technological and social factors eliminated the necessity for domestic gardens, if not their desirability. The invention of the canning and later the freezing of vegetables; the use of speedy motor transport for the consumer and the producer-shipper of both flowers and vegetables; and the disappearance of the domestic carriage horse and other animals, with their useful wastes, all tended to make the domestic garden a discretionary space, not a necessary space. We are reminded at garden meetings that "statistics show that gardening is America's most popular outdoor activity" or some such statement. Perhaps. But the popularity of gardening does not imply the continuity of the garden as a central social space dictating its own obligations and relationships any more than the popularity of television-watching as our favorite indoor activity implies continuity of the family hearth.[8] The recent academic interest in the continued centrality of the garden among immigrant and ethnic groups (Giraud 1987, Warner 1987) only highlights the decline of the garden as a work and social focus for most of us.

As the folk garden, a base for the high art, has declined in this century, so has that high form itself. The obvious absence of cheap skilled labor and the disappearance of the professional gardener in much of modern society seem too easily ignored in our current concern over bringing back garden meaning, although we are dramatically reminded on our summer tours when we see the decay of a Moghul showplace or Jekyll masterpiece. Lawn maintenance services with their mower and sprayer-laden pickup trucks have replaced the journeyman worker who was the central character of every English garden between the world wars (e.g., Blythe 1969), just as the rental plantscape industry has replaced the gardeners and apprentices who once groomed the grounds of every elegant hotel. These restraints were better recognized decades ago by the pioneers of contemporary American landscape design, and in fact served as a major argument for their advocacy of a wholly new kind of "modern outdoor living." In the landscape, as in so many parts of modern life, speed and standardization have replaced patience and skill, and industry has replaced craft.

But it is debatable whether the garden as art or place of meaning has ever been much accepted or understood in America. Moore and Mitchell (1986) have given us a catalog of believable reasons why garden art has not been important in our country, reasons that include the prevalence of space and its use instead of walls as a mark of status; a fascination with converting or ordering that space on the grandest scale (e.g., their image of the midwestern township system as a great gridded Moghul garden); an addiction to movement; and concentration on our "unbeatable natural wonders." Other writers confirm or expand these contentions. George Stewart (1954) pointed out how constant movement has shaped our tastes in food and drink and, by my inference, the garden. Duncan's (1973) description of landscape among the Westchester County gentry emphasized the role of modesty, extensive land, and a shaggy, almost shabby rusticity, instead of formal arrangement and display, as an indicator of status. Lowenthal's (1968) identification of three prime images of the American landscape as size, wildness, and formlessness could well stand as a definition of anti-garden. Elbert Peets (1927, p. 100) said it succinctly over sixty years ago: "The current American is not a gardener . . . he does not care for plants as plants, though he loves grass, trees, sunlight and panoramic views as much as ever." The English bequeathed us a taste for lawns. We used them to unify our free-standing houses in one of the grandest developments of civic design that history shall know. But their taste for gardens seems not to have crossed the Atlantic. The guidebooks tell us that crowds make summer Sunday visits to Sissinghurst impossible. How many readers of the *New York Times* garden section have heard of Fletcher Steele or Russell Page, or could put a name to those familiar pictures of the Donnell garden?

But if Americans, generally, have neglected the garden, there came a time when American designers, specifically, in some sense actively turned against it. The relationship between the garden and the ascendancy of contemporary landscape design from revolution to establishment dogma is a complex subject, rife with ambivalence, crying for documentation, analysis, and interpretation. Reaction against the grip of tradition, particularly English tradition, on our landscape design and education can be traced in a series of articles that began with Peets (1927), was articulated by Rose (1938, 1939a, 1939b, 1939c, 1939d) and Eckbo, Kiley, and Rose (1939a, 1939b, 1940),[9] and both reprised and rephrased by Streatfield (1970) three decades later. None of these writings attacked the garden *per se:* Rose's manifestos carried the title "Freedom in the Garden" and Streatfield's title, "The Tyranny of the Garden," is an inaccurate, if catchy, summary of his message. To Peets (1927, p. 94), the enemy was the English informal landscape and its priesthood "wallowing in the pool of nature sentiment"; he explicitly extolled both the Italian formal garden and the folk garden. Rose (1938, p. 643) briefly dismissed the English landscape style ("'Informal' came to mean 'formless'") and concentrated his attack on the formal and axial *Beaux Arts* tradition. Eckbo et al. (1940) advocated larger areas of concern: the urban, rural, and primeval environments. Streatfield (1970) characterized the limitations of the visual and asked for more attention to ecological and behavioral approaches. Church (1955) and Rose (1958) addressed their first books to garden design and included the word in their titles. Whatever their polemics for new models, new philosophies, and larger arenas for design, their real world staple was the domestic garden; the Donnell garden stands to American landscape design as the Villa Savoie does to International Style architecture. If by garden we mean an outdoor living space as an extension of the house, then the garden continued as the rock on which landscape architectural practice and education were based, a role equivalent to that of the single family house in the rise of modern architecture.

But if by garden we mean the archetype of nature formed for meaning, if we mean the carrier of a culture's central symbols, it is hard to escape a sense of the garden's devaluation, a sense of it as the only available commission, executed in preparation for the awaited entry into the larger landscape and social relevance. Eckbo et al. (1939a, 1939b, 1940) completely omitted gardens from their manifestos of landscape concerns in the *Architectural Record* articles. Peets (1927, p. 100), in his frequently cited but seldom read article, concentrated his hopes for the future on dams and stadiums, on "concrete mixers and steam shovels . . . booster energy and national highways." Landscape educators might also ask themselves when the course on "Garden Design" disappeared or was relegated to a non-professional, community-oriented service course, or when that introductory site design problem, the "Smith Garden," became the "Smith Residence," or when and why "plants" became "plant materials." Somehow the garden came to symbolize the dead hand of tradition, a rich man's toy, an emblem of social irrelevance or outright oppression. Can we imagine a post-revolutionary designer designing a space for outdoor living like Gatsby's, where in "blue gardens, men and girls came and went like moths, among the whisperings and the champagne and the stars" (Fitzgerald 1963, p. 49)? In their search for larger scale, bigger issues, and greater impact they turned to models such as TVA, parkways, and plans for river basins. Although this was a realistic and socially laudable redirection, we could also wonder whether a male-dominated profession thought that ladies grow flowers, but real men realign rivers.

The American landscape designers' rejection of the garden was supplemented by an earlier, broader, and more influential rejection, that of the mainstream European modern architectural tradition. The attitude of the classic modern architects to the landscape is an important subject just now beginning to receive the attention it deserves. It is a subject of great difficulty, because those attitudes were usually implicit not explicit.[10] One undocumented and very simple characterization can be offered: the classic modern tradition rejected the garden and enshrined the park. Kent, we are told, leaped the wall and found all England a garden. The modern movement breached the city walls, dispersed the city itself, and envisioned all nature as a park. That park was an English landscape garden grown global and simple. The vision was buttressed by the nineteenth-century movements of social reform and municipal hygiene, and by the garden city. Nature was to be a healthy and utilitarian, but passive, redemptive setting. Visually it was to be as clean, simple, and uncluttered as modern buildings were compared to their Victorian predecessors. It was to be a great pastoral sward upon which point blocks, borne by *pilotis*, rested as lightly as spaceships, as tastefully as cattle in a Repton Redbook. It was the vision of Ville Radieuse, of Roehampton, and of the high-style corporate office parks lining our freeways.

The difference between garden and park useful for this discussion is not one of scale, or user, or even necessarily uses, but of the role of nature. The essence of garden is not only control of nature but demonstration of that control. Such demonstration is inessential, perhaps even distracting, in the park. The park lacks concentration on both the grown, the product of control, and the growing, the process of control.[11] This is a non-trivial difference, as Francis (1985) has shown in his study of the perceptions and motivations of park users compared to those of community gardeners. The park, as realized in the great Anglo-American visions of the nineteenth century, and in Ville Radieuse and classic modern architecture, conceptualizes nature as a benign and supportive environment conducive to moral activities and attitudes. This park is the idealized pastoral realized, a harmonious natural setting supporting and enhancing harmonious human activities. That the gardens of the nobility became the parks of the people is a reference not only to urban history but to designers' dreams. Control of nature is an end in itself in the garden, a means to human ends in the park. The garden art fails when it becomes so obviously contrived as to get in the way of meaning. The park fails when the benign becomes the bland, and no meaning at all is conveyed. That has been the fate of what began as a grand vision of modernism.

That vision, however, is still alive. It is a vision of optimism and liberalism, of friendly but too often banal nature. It is a vision that in a mindless form has pervaded our consciousness and our landscape. The landscape as debased park has produced a world we find troublesome and even threatening, a world so extensive that wild nature has become relief and haven. Garden, become park, become landscape, has reversed the traditional relation of garden to wild. The wilderness has become an inverted, particularly American, *Hortus conclusus*, fenced off for the pack-framed pure of heart.

The Resurgence of the Garden

The reasons for resurgence of the garden might be as obvious, once articulated, as the reasons for its decline. First, we are seeing a reaction against the uniform, universal landscape of the classic moderns, for whom the preservation of global diversity was hardly a major worry. The international landscape has turned out to be not only pastoralized, but homogenized. Whether this loss of diversity is real or simply a new, finer-grained pattern of diversity, self-conscious and nonregional, is a question I have discussed elsewhere (Riley 1980) and will only note here. Equally suspect is the common charge that the universal landscape produces "placelessness" now that "place" has become the catch phrase that "function" was to our predecessors. Garden styles have been as international and uniform (e.g., the bedding plants of British colonialism) as the building styles of European colonialism or workers' democracies. The garden is now seen as an antidote to the vaguely corporate, universal landscape that bores and bothers us. We could also view the return to the garden as one more manifestation of the privatization of our society, the garden replacing the park just as pool and patio replaced the porch. Is the garden as fitting a symbol of Reaganism as Greenbelt and then Columbia, Maryland, were for the naive confidence of the Roosevelt and then Kennedy-Johnson eras?

Simple cycles of fashion probably play a part in the resurgence of the garden as well. Among landscape architects, renewed interest in the garden might be seen as a predictable return to the historical, identity-bound core of the profession, redressing the imbalance attendant upon the expansion of social concerns and professional boundaries of the preceding two decades. Architects and the architectural press have also been conspicuous in the return to the garden and simple architectural conceit might explain much of their attention and rhetoric. From the earliest examples of the English landscape style through both European and American classic modern architecture, from Rousham through Villa Savoie, to the Donnell garden and the Farnsworth house, the organizing geometry of the house stopped at its walls. Architects' recent fascination with the garden might have been forecast as part of the recurring cycle of building retreat and then advance in the continuing naturalist vs. formalist landscape debate. Current design rhetoric is similar to that of the two great turn-of-the-century protagonists of the formal garden, Blomfield (Blomfield and Thomas 1892) and Sitwell (1909). Both writers faulted the English landscape style that dominated eighteenth- and nineteenth-century garden design for its failure to relate the landscape to the house it served. The comparison hardly justifies some contemporary architectural silliness, but it provides a respectable precedent for the garden grids of Charles Moore (Moore and Mitchell 1983) and Barbara Stauffacher Solomon (1982). It is fun, if unprofitable, to speculate whether the impact of Moore's clever, articulate writings and Solomon's gorgeous graphics played any part in the renewed prominence of gardens in landscape architectural publishing. What matters is that designers are back in the garden, forming nature and seeking meaning.

The Search for Meaning: Problems

There are many problems, some obvious but some less so, in restoring garden meaning, obstacles to once again centering powerful and accepted symbolism on the garden. Many of these result from new attitudes towards nature. Control over nature, the essence of garden meaning, has become so commonplace as to mean little. Sixty years ago Peets (1927) looked to vast highway systems and dams for the act of dramatically molding nature. Those demonstrations first became routine, and then commonly questioned for their environmental effects and philosophy. The Victorian era was the last great age of gardens in the European tradition. If we today denigrate the quality and philosophy of nineteenth-century design, we can only wistfully wonder at the fascination with plant display that characterized that time, the preoccupation that Nicolette Scourse (1983) has described in detail in her book, *The Victorians and Their Flowers*. There might be no generalizable answer to the question of why one culture places profound meaning on plants and gardens and why another does not, but we can see how many aspects of Victorian society make their predilection understandable. The Victorians reveled in technological process and virtuosity, much of it centered on construction and environmental control. Improved, less expensive glass production, cast-iron framing, and advances in economical, easily available heating and hydraulic systems in artifacts ranging from Wardian cases to huge conservatories, when combined with the English love of plants and gardening, opened new worlds of botanical control and display. If we add to that the imperialist sweep of exploration, colonization, and collecting, then the Victorian preoccupation with gardens (and zoos) as social focus and value carrier seems not only logical but almost inevitable.

No such obvious technological and social synergism advances the role of the garden today. Our technological advances and our fascination are focused on personal and portable electronic wonders relating little to mastery of the natural environment (although much to the way we enjoy that environment), advances beyond the intuitive understanding that could comprehend, but still wonder at, the conservatory at Chatsworth.[12] Easy acceptance of and then growing dismay over our power to rework nature, combined with a traditional American preference for remote and spectacular grandeurs, have redirected our search for symbolic nature from the garden to the wilderness.

What we ask of nature, and how we use it, have changed as well. Electronics and personal transportation have produced new ways of enjoying the natural world. We emphasize motion and activity, individual, transitory experience, often heightened by sophisticated technology. Our mobility and technology allow us not only to see many places briefly, rather than lingering in and deeply understanding only one, but to experience nature in new ways. Dirt-biking, hang-gliding, and surfing, often abetted by personal cassette players, emphasize not lingering or traditional social meanings but brief, intense, individual sensate experience, often tinged with risk or even danger. Jackson (1957) characterized this new relation with nature as the abstract world of the hot rodder. I have (Riley 1980) described it elsewhere, less elegantly, as the high-tech, hedonistic quick-fix. Whatever we have gained, we have lost the taste for those quiet times of contemplation central to the experience of so many older gardens.

Shared symbolism, so central to the high art of the garden, is also a difficult problem for our time. The attempts of modern designers to achieve it show a history of failures, from Mies's steel-on-concrete-on-steel corners, through Venturi's gilded antenna, to Moore's plated capitols. The symbols that have empowered great architecture and great garden art are what Mary Douglas (1970) called condensed symbols, symbols that carry not just one meaning but accretions of many meanings, layered upon each other and over time. They are symbols that are commonly agreed upon, not designer-chosen, that connote deep affective meaning, not quick cleverness, and that are integral to a context that is culturally agreed upon as appropriate. Are there such symbols for our time, or do we lack them, at least as they are capable of being expressed in physical design, altogether? Rapoport (1982) has suggested that such powerful symbols as we have are more likely to be individual than communal. This seems plausible for a culture as centrifugal, as pluralistic, as individualistic, and as privatized as ours. Such a lack of shared symbolism does not rule out the garden as a carrier of powerful meaning but it does discount the likelihood of meanings that speak strongly to the whole society.

The Search for Meaning: A Program

From where might such meaning come? How might we try to build into our gardens a symbolic content to match that resurgence of design interest among designers? Such meaning can come only from a careful, serious assemblage of scholarship, research, and design, a complex of work that follows a program, not a fashionable whim. Our scholarship should analyze, synthesize, and interpret garden history in order to pose questions for research. Our research should pose and test hypotheses that might help answer those questions. Our design should both apply the results of our research and serve as a laboratory for it. This process, of course, should be circular, not just linear.

SCHOLARSHIP

The scholarship that will help us is a far cry from the current deluge of picture books and from the model of garden history that delves ever more deeply and narrowly into archival minutiae. We need to begin with interpretive and rigorous exploration that goes beyond the garden and its great designers to explore the relation of garden design to its political, social, economic, and technological context. Many histories and a few journal articles offer token recognition and simplistic, time-worn generalizations on these topics, but few take it as their central concern. Those works that deal seriously with garden meaning in history relate it to its sister arts and intellectual simplifications about the "spirit of an age." One does not need to be a Marxist, or even a cultural materialist, to realize that ideas do not arise completely independently of the economic and social matrix of their time. Essays by Fabricant (1979) and Williams (1987) have broadened our understanding of the context of the Georgian garden. Scourse's (1983) book on Victorian flowers is a model for comprehensive, integrated treatment of the social, economic, and technological setting upon which garden art is based. Such studies will provide no easy answers for contemporary design. The scholarship we need will not tell us what a garden means or what a garden should mean. It can suggest why gardens were or were not a locus of meaning at particular times and why and how that meaning emerged from the society that produced it. Such work will serve as a catalyst for designers, a stimulus to thinking seriously and reflectively about meaning, a standard against which a designer's clarity and explicitness about assumed meaning and its relation to society can be measured.

RESEARCH

A world of potential research into garden meaning lies before us, a world as yet hardly explored, let alone mapped. The traditional literature is filled with assumptions about the satisfactions gained from gardens and gardening. These assumptions are often phrased in syrupy language, but are at least explicit. Nan Fairbrother's (1956) chapter "Why Men Have Gardened" alone contains enough clear and succinct statements about gardening as an activity and the garden as a place to keep researchers busy for years. Environmental psychology possesses a repertoire of potentially useful concepts for structuring and exploring these hypotheses: competency theory, cognitive structuring, information processing, play theory, and so on. Joachim Wohlwill (1983) and Roger Ulrich (1983) have laid out clear exemplars for the larger world of satisfaction from nature in general. The same must be done for the smaller, more intense world of the garden. Fairbrother's (1956, p. 252) description of gardening as a tranquil occupation, "busying the mind with quietness," illustrates· Rachel Kaplan's (1973) concept of involuntary attention. Yeats's classic statement of the pastoral garden, "The Lake Isle of Innisfree," bears a similar relation to Wohlwill's (1983) exploration of nature as stimulus and to Stephen Kaplan's (1983) concept of restorative environment. What psychological concepts or processes might help us understand or better categorize Sun's (1987) intuitive concept, referred to earlier, of nature's garden, the artist's garden, and the poet's garden? The testing of traditionally assumed garden satisfactions with the concepts of techniques of environmental psychology is a research direction as fruitful as it is obvious. That it has hardly been suggested tells us much about the superficial thinking of designers and the constantly lamented gap between designers and social scientists.

Meaning in the garden, that response beyond the sensory and pictorial, that essence of the poet's garden, has never been confined within the garden itself. Garden meaning has been shared with other art forms and landscape forms. Some designers are conducting their own search for meaning without waiting for the scholarship or research of others. They are searching both inside and outside the garden.

Designers are turning to several sources for help in this search for meaning. Contemporary high art, with its post-modern philosophies and forms, is one of these. This is reasonable; the link between the garden and other art forms, particularly sculpture, is traditional. Almost any contemporary garden is sure to contain a piece of equally modern sculpture, but integration of garden and art is more than the obligatory gesture towards sculpture, more than pastoral plop art. Martha Schwartz and Ian Hamilton Finlay (Abrioux 1985) have in different ways moved beyond the garden as a green backdrop for isolated sculptures to serious explorations into the synthesis of sculpture and garden. Even the most thoughtful synthesis of contemporary plastic and garden arts, however, poses a problem that must be faced. The tenets and vocabulary of contemporary art are not shared among the wider society. A simple fact that is probably regrettable, but surely true. The problem is not elitism. The latter is a charge often flung at garden designers by the too few socially conscious designer-activists; it is a charge that mainstream designers seem proud to accept. The function of art, after all, is to lead public taste, not to pander to it. But the proper definition of elitism is rule or domination by a powerful clique, by the best and the brightest. Leadership is fine. But if designers are to accept the charge of elite leadership, they should look back once in a while to see whether anyone is following. Contemporary art and avant garde design are not elitist, but arcane. The problem today is compounded by the fact that by the time a style is understood and absorbed by a wider public, the designers have moved on to other publication-rewarded themes. These remarks are simplistic, but we ignore such simple reminders constantly. These difficulties do not negate the potential of contemporary garden/high art synthesis, which can and will be a powerful experience. They do sketch out limitations. The problem is similar to that of gardens based on personal symbolism: meaning is shared by few, not many. The difference is that in a garden built on personal symbolism we do not know who will share the meanings. (A research program such as I have outlined would, however, provide clues.) In a high art garden we do know who will share meaning: the designers' peers and readers of the arts literature. The argument that shock is good for you might be valid but it is not sufficient. Shock might prepare one for meaning; it does not produce it. (That tenet of certain other contemporary architectural gardens, that schlock is good for you, need not be taken seriously.)

Earth art, a clumsy term covering the work of diverse environmental artists using natural materials and phenomena, shares many of the beliefs and vocabularies of the garden as high art. For our exploration of garden meaning, however, it can be better understood as being opposite. If the high-art approach conveys meaning to a small subculture, earth art attempts to convey universal meanings. It is precultural in its return to immediate sensations and the primal awe of experiencing nature and its transformations, those feelings thought to be universal from our beginnings as hominids. It is an attempt to reanimate our art with the power of the sacred grove.[13] In a pre-garden society, however, awe and symbolism were constants in the human-nature relationship. The great archetypal environmental artworks—Stonehenge, Avebury, Silbury Hill—did not create that awe; they focused and celebrated it. Current earthworks are not foci amidst such continuity, but isolated incidents amongst a culture bent on different ends. The distance between our sophistication and the naiveté Henry Adams found in the Virgin worship of Chartres is as nothing compared to the distance between ourselves and the builders of Silbury. Can the meaning sought by Lucy Lippard (1983) and others move beyond the transient experience, even survive, in a culture that is at best indifferent to it? Some of the radical feminists who seek to establish religious ritual around Mother Earth's moundings seem to be the only ones addressing this problem. The power of many contemporary earthworks is formidable. Are they gardens? More pertinently, do they fulfill the role of gardens? If any work of art that forms nature to provide meaning is a garden, then they are gardens. If gardens are expected to move beyond primal response and offer accretion of layered meanings predictably shared among a wide audience, the answer is problematic. Earth art is best understood as a pre-garden, an attempt to short circuit all the centuries of shared symbolism turned empty, to return to the time when nature *was* Garden, maybe slowly to build again to nature *as* garden.

Other designers have turned to a very different reference in their search for meaning—the regional landscape. Warren Byrd's (1986) designs are based on the evolving landscape of tidewater Virginia, and Terry Harkness's (1986) on the agricultural landscape of central Illinois. Scholars are only beginning to explore the relationship between agriculture and the

Restoring the illimitable: the work of Bernard Lassus. (Drawing courtesy of Bernard Lassus, from original publication)

birth of garden art. What we do know is that the larger landscape, both cultural and natural, has been at the base of some great garden styles. The Chinese tradition is the epitome. It was the result of layer upon layer of abstractions of the larger landscape: from landscape, to landscape myth, to landscape poem and painting, to garden. It would take years to move from Byrd's and Harkness's start to that kind of art (years being the electronic age equivalent of earlier dynasties), but the possibility is there. The meaning source here is widely accessible; if it is limited by region it cuts across lines of class and taste. The connoisseurs of silos and hedgerows probably outnumber the connoisseurs of New York art, at least in central Illinois. Their lore is accessible, if one is willing to substitute a trip to the courthouse café for an evening with the *New York Review of Books*.

One last approach to garden meaning, the most provocative of all, is worth exploring. It is seen in the work of Bernard Lassus (1983), the most innovative garden thinker of our time. Lassus seeks to rebuild garden meaning by turning not to sources beyond the garden, or before the garden, but to the essence of garden itself. He sees this essence as the contrast between the knowable and the illimitable, between the mappable and tangible world of the garden and the *terra incognita* of the world outside the garden. In his view the failure of garden meaning in our time comes from the fact that our entire planet is now not only knowable, but known. There are no blank places on our maps or in our minds, no heroes where there could be dragons. The power of garden depends on the power of its opposite. His gardens become true gardens again because he supplies that illimitable. Our horizontal world is too well known so Lassus provides us with the vertical: bottomless voids into which dropped pebbles travel forever. He builds caverns for us and engulfed villages that are seen and then vanish, and giant insects emerging from the water. The artifice of Lassus would destroy our complacency by providing uncontrolled nature. He brings back fear of nature, to the benefit of the garden. (Maybe along with fear, we should bring back sex, which also seems to have been missing from the garden for too long.) If his conceptual approach is unique, Lassus also stands almost alone among contemporary designers in welcoming technology in the service of experience. He uses it to hint at fear, at the tingle of the beyond and the awesome upon which the older vision of the sublime rested. What a garden a team of Lassus, Vicino Orsini, and the technical wizards of Disney Enterprises might have made!

Lassus also explicitly brings fantasy back to the garden. It has never been gone, just ignored. Or maybe looked at from the wrong end. Gardens are commonly thought of as fantasy

Hestercombe. (Photograph by Robert B. Riley, courtesy Bob Riley Landscape Architecture Collection, Digital Collections at the University of Illinois Urbana-Champaign Library)

concretized. At its most serious this view sees the garden as a prime, tangible expression of a culture's most powerful fantasy—its world view. At the other extreme the garden is seen as the place to indulge in the concretization of fantasy turned fluff. Architecture is serious business in this view, but the garden is where we can build our whimsy. It is significant that Charles Moore, the only American architect of our time who unashamedly deals with fantasy and whimsy, is leading the architects' march back to the garden. The trip to concretized fantasy is a dangerous one. Moore's discipline and intelligence keep his work this side of silliness. When such checks are removed we get Stanley Tigerman's terminally cute topiary. But the garden can be seen as not a concretization of fantasy, but a spur to fantasy. The Georgian arcady was a fantasy cue for the *cognoscenti*. Disney World is a concretized fantasy, but can we doubt that most of its popularity lies not in the literary fantasies it programs but in the internal fantasies it releases, fantasies that are, in the jargon, both interactive and open-ended?

That the garden deals with death is only one of the two hallmarks that distinguish it from architecture. The other is that it is the unprogrammed space *par excellence*, the space that serves the dwellers' unpredictable, internal activity, not designer-codified external activity. Its power to generate fantasy, both shared and individual, lies in the fact that it is an artificial place made from the stuff of the real. It is real enough to be convincing, but not so real as to constrain the unreal. That is a powerful form of meaning, or the means to it.

Our search for garden meaning raises the problem, already alluded to, of audience. When a garden speaks of meaning shared by a wide community, several things happen. Most simply, satisfaction is gained by more people, as it should be in a democratic society. Beyond that, it allows communal sharing and hence reinforcement and enhancement of those joys. The garden is linked to cultural values, social forms, and the other arts, reinforcing them and in turn being reinforced. Lastly, it assures its own survival by receiving social approval, by becoming important. Survival does not ensure quality but it is a prerequisite for the development of quality. Gardens need consumers. Garden meaning requires more than an adequately broad audience. It requires connoisseurship. Meaning and connoisseurship are inextricably linked. A critical mass of critics, an informed, articulate group of garden participants and patrons, enriches garden meaning and art in two ways. First, it demands ever more perfection and innovation in the garden and rewards the designer who provides them. Second, such patronage permits accretion, the building of meaning upon meaning. This accretion is a mark of great and enduring art forms. It rewards understanding with joy at many levels, from the sensory to the cerebral, from the literal to the abstract.

Connoisseurship rests on literacy. The literalness required of a garden is in inverse proportion to the literacy of its audience. The long decline in the English garden from Georgian to Victorian times can be read as the withering of connoisseurship and the rise of consumerism. The greatness of the Chinese garden can be read as a long progression from the literalness of the landscape as itself to the layered references and abstraction of the Wang Shi as landscape of the mind. This is a way to understand both the problem and the potential of Byrd's and Harkness's gardens. One or two gardens will be interesting, but if too literal in their references they will soon lose their return and be discovered once and for all. To test their potential we need many of them, with varying referrants and with interplay between varying levels of abstraction from those referrants. The connoisseurs' demands for ever more complex and intricate abstract relationships can corrupt a style as surely as descent into literalness in the service of sentiment. It will be a long time before we need face those worries, however. Garden connoisseurship has died just as has the garden craft. Sitwell's (1909) *On The Making of Gardens* is the last great work of that tradition, its zenith, and its climax. To read it is to understand the height of intellectual power and emotional richness that connoisseurship reached, is to understand the critical role of the classical education and the grand tour upon which that connoisseurship rested, is to understand the loss we have suffered. Rebuilding garden connoisseurship will be even more difficult than rediscovering the garden craft. Whether the short cycles and intense media exploitation of the contemporary design world will hinder or help its development is problematic. So is the question of which comes first, consumers or connoisseurs. What is certain is that we need both.

In Conclusion

A few simple rules will help in our search for meaning.

We should remember that we live in a diverse, pluralistic culture, in which widely agreed-upon meaning and powerful, accepted symbols are the exception. Because of this we need to be explicit about what meanings we hope to convey, and why, and to whom. We need to identify explicitly our consumers and our connoisseurs and to distinguish between them.

We should acknowledge technology, not flee from it. There is no more reason for us to ignore computers and holograms, light shows and lasers, in our gardens than there was for the Victorians to ignore hydraulic and heating technologies in theirs.

We should remember that a garden is not a building and to emphasize the difference. A garden is about birth but equally about dying. Its power depends upon this. A garden is not just a fragrant piece of architecture. A garden is about fantasy. A garden can be not only a fantasy realized, but a fantasy provoked.

We should accept that this is a time of exploration, not consolidation. We might even rejoice in that fact. But we should remember that exploration is not aimless wandering. We need to think and to reflect, and to avoid the symbols sweepstakes. Our scholarship should shape our research, our research should shape our design. But that is not enough. If we need research in the service of design, we can also exploit design in the service of research. Each of our garden designs could pose a hypothesis and a plan to test it. That is the difference between exploration and wandering.

The path to contemporary garden meaning will be a hard one, as lined with cul-de-sacs, contrived vistas, and artifice as the garden itself. It is a path worth taking. It is a path that will bring disappointments as well as pleasures. We do not know where it will lead us. That is as it should be and as it must be. As Don Juan teaches us, there are no destinations. There are only journeys.

NOTES

This essay is based in large part upon a study supported by a National Endowment for the Arts Fellowship. I am also grateful to Anne Whiston Spirn for her many perceptive comments, questions, and suggestions.

1. My use of the word "man," and not a gender-free substitute, is intentional. Both Ortega y Gasset (1972) and Shepard (1959, 1973) assume, the former implicitly but the latter explicitly, that hunting was originally the exclusive domain of males. That is a common assumption and one that has probably elevated the status of the hunt. The common parallel assumption, that garden is woman's domain, has done little to elevate the status of the garden, however.

2. Stephen Kaplan (1983) distinguishes among three related concepts often subsumed under the term "control" in people's relation to the environment. These are the assurance of non-randomness, or things being "under control"; participation, or things being "not beyond one's control"; and effective personal power, or being "in control." All three help in understanding the diverse meanings of a garden.

3. Editor's note: This essay first appeared in a special issue of *Landscape Journal*, guest edited by Anne Whiston Spirn, *Nature, Form, and Meaning* (1988).

4. It is not obvious how a non-human material "humanizes" an environment, or why certain non-human materials such as plants are thought to accomplish this while others, like rats or weeds or bugs, do not. Maybe Fairbrother (1956) is using a concept sloppily, just as architects prattle endlessly about "human scale." But she is expressing a common belief and one worthy of research.

5. This definition of the garden as control over nature for human pleasure is useful and concise but raises some perplexing questions. By nature do we mean *living* nature or will rock and water do? Does control include imitation of the natural by the artificial? Is an atrium filled with plastic plants a garden? Does control ultimately include total banishment? There are no plants in Ryoanji but we call it a garden. Is it a garden because of its relation to the trees visible beyond? Is it a garden because its rocks and gravel *refer* to nature? Is it then not a garden or is it the final garden, a reductionist key to the ultimate garden in and of the mind?

6. When we refer to the basic garden element we assume plants are meant. Karel Capek (1984), the Czech author best known for coining the word robot, has written a witty, dissenting essay in which he suggests that the richest and most fascinating element in the garden is not the plant but soil, and provides a catalog of its attractions.

7. This is a quotation as common these days as Bacon's reference to the Lord God and the first garden once was, and just about as useful for actual garden design work.

8. I will leave for others the question of whether the "yard" or "lawn" (terms etymologically distinct but almost interchangeable in contemporary usage), so important in the American environment, should be considered a garden.

9. The similarities and differences between these calls for a new approach and the earlier arguments of Jens Jensen and Wilhelm Miller for a style based on the quality of the American landscape and not European precedents are worth study.

10. Frank Lloyd Wright and the European modernists represent distinct, even polar, approaches to the building-landscape relation, but there were also distinct visions within the European tradition itself. Alvar Aalto surely would have to be distinguished from the so-called International Style mainstream. Even within the latter, Le Corbusier and Hilberseimer held different ideas of the role of nature and the landscape.

11. This distinction between park and garden based on demonstration of control, like the earlier definition of garden based upon exertion of control, is handy and concise but far from watertight. Where, for example, does a "garden" by A. E. Bye fall under such a dichotomy? And how do we deal with Sylvia Crowe's contention that the royal hunting park was one of the two archetypes for the concept of garden?

12. The *reductio ad absurdum* of computerized, electronic miniaturization of nature will be a solar-powered Japanese wristwatch garden, custom-programmable for owner-chosen displays of growth and seasonal change and imported by Hammacher-Schlemmer.

13. The sacred grove and Disney World can serve as antipodal limits to the garden, and to human relations with nature, in several ways. They can be viewed, for example, as pre-garden versus post-garden, or as the primacy of meaning versus its trivialization, or as nature's magic versus human technology masquerading as nature's magic.

REFERENCES

Abrioux, Yves. 1985. *Ian Hamilton Finlay: A Visual Primer.* Edinburgh: Reaktion Books.

Blomfield, Reginald, and F. Inigo Thomas. 1892. *The Formal Garden in England.* London: Macmillan Publishing Co., Inc.

Blythe, Ronald. 1969. *Akenfield: Portrait of an English Village.* New York: Delta Books.

Byrd, Warren T., Jr. 1986. "Tidal Garden: Eastern Shore of Virginia." *Places* 3, no. 3: 22–25.

Capek, Karel. 1984. *The Gardener's Year.* Madison: University of Wisconsin Press.

Church, Thomas. 1955. *Gardens Are for People.* New York: Van Nostrand Reinhold Co., Inc.

Douglas, Mary. 1970. *Natural Symbols.* New York: Pantheon Books.

Douglas, William o. 1961. "Wilderness and the Molding of American Character." In *Wilderness: America's Living Heritage*, ed. David Brower. San Francisco: Sierra Club.

Duncan, James. 1973. "Landscape Taste as a Symbol of Group Identity." *Geographical Review* 63, July: 334–55.

Eckbo, Garrett, Daniel U. Kiley, and James C. Rose. 1940. "Landscape Design in the Primeval Environment." *Architectural Record* 86, February: 74–79.

———. 1939a. "Landscape Design: The Urban Environment." *Architectural Record* 85, May: 70–77.

———. 1939b. "Landscape Design: The Rural Environment." *Architectural Record* 87, August: 68–74.
Fabricant, Carol. 1979. "Binding and Dressing Nature's Loose Tresses: The Ideology of Augustan Landscape Design." In *Studies in Eighteenth Century Culture,* ed. Roseana Rund. Madison: University of Wisconsin Press.
Fairbrother, Nan. 1956. *Men and Gardens.* New York: Alfred A. Knopf, Inc.
Fitzgerald, F. Scott. 1963. *The Great Gatsby.* London: Bodley Head.
Francis, Mark. 1985. *The Park and the Garden in the City.* Davis: Center for Design Research, University of California.
Giraud, Deborah D. 1987. "The Meaning of Gardens to Hmong Refugees." In *Meaning in the Garden: Proceedings of a Working Conference Held to Explore the Social, Psychological, and Cultural Dimensions of Gardens,* ed. Mark Francis and Randolph T. Hester, Jr. Davis: Center for Design Research, University of California.
Harbison, Robert. 1977. "Green Dreams." In *Eccentric Spaces.* New York: Alfred A. Knopf, Inc.
Harkness, Terence. 1986. "An East Central Illinois Garden: A Regional Garden." *Places* 3, no. 3: 6–9.
Healy, Vince. 1987. "The Hospice Garden: The Visitor and the Grieving Process." In *Meaning in the Garden: Proceedings of a Working Conference Held to Explore the Social, Psychological, and Cultural Dimensions of Gardens,* ed. Mark Francis and Randolph T. Hester, Jr. Davis: Center for Design Research, University of California.
Helphand, Kenneth. 1985. "The Garden." Paper given at the meeting of the Council of Educators in Landscape Architecture, Urbana, Illinois.
Howett, Catherine. 1987. "Gardens Are Good Places for Dying." In *Meaning in the Garden: Proceedings of a Working Conference Held to Explore the Social, Psychological, and Cultural Dimensions of Gardens*, ed. Mark Francis and Randolph T. Hester, Jr. Davis: Center for Design Research, University of California.
Hudson, W. H. 1942. "A Boy's Animism." *Far Away and Long Ago.* New York: E. P. Dutton.
Jackson, J. B. 1980. *The Necessity for Ruins and Other Essays.* Amherst: University of Massachusetts Press.
———. 1957. "The Abstract World of the Hot Rodder." *Landscape* 7, no. 2: 22–27.
Johnson, Hugh. 1979. *The Principles of Gardening.* New York: Simon & Schuster, Inc.
Kaplan, Rachel. 1973. "Some Psychological Benefits of Gardening." *Environment and Behavior* 5, no. 2: 145–62.
Kaplan, Stephen. 1983. "A Model of Person-Environment Compatibility." *Environment and Behavior* 15, no. 3: 311–32.
Lassus, Bernard. 1983. "The Landscape Approach of Bernard Lassus." *Journal of Garden History* 3, no. 2: 79–107.
Leopold, Aldo. 1964. *A Sand County Almanac.* New York: Oxford University Press.
Lippard, Lucy. 1983. "Feminism and Prehistory." In *Overlay.* New York: Pantheon Books.
Lowenthal, David. 1968. "The American Scene." *Geographical Review* 58, January: 61–88.
Moore, Charles, and William Mitchell. 1986. "American Edens." *Places* 3, no. 3: 60–64.
———. 1983. "On Gardens." Mimar 8: 23–29.
Olin, Laurie. 1986. "Brillig and Contrary Gardens." *Places* 3, no. 3: 52–55.
Ortega y Gasset, José. 1972. *Meditations on Hunting.* New York: Charles Scribner & Sons.
Peets, Elbert. 1927. "The Landscape Priesthood." *American Mercury* 10, no. 37: 94–100.
Rapoport, Amos. 1982. *The Meaning of the Built Environment: A Nonverbal Communication Approach.* Beverly Hills, CA: Sage Publications, Inc.
Riley, Robert B. 1987. "Flowers, Power and Sex: Themes in the Literature of Garden Meaning." Paper given at the conference *Meaning in the Garden,* University of California, Davis.

———. 1980. "Speculations on the New American Landscapes." *Landscape* 24, no. 3: 1–9.
Rose, James. 1958. *Creative Gardens.* New York: Van Nostrand Reinhold Co., Inc.
———. 1939a. "Articulate Form in Landscape Design." *Pencil Points* 20, February: 98–100.
———. 1939b. "Plant Forms and Space." *Pencil Points* 20, April: 227–30.
———. 1939c. "Landscape Models." *Pencil Points* 20, July: 438–44.
———. 1939d. "Why Not Try Science?" *Pencil Points* 20, December: 777–79.
———. 1938. "Freedom in the Garden." *Pencil Points* 19, October: 640–44.
Sassoon, Siegfried. 1949. "Villa d'Este Gardens." *Collected Poems.* New York: Viking Press.
Scourse, Nicolette. 1983. *The Victorians and Their Flowers.* Portland, OR: Timber Press.
Shepard, Paul. 1973. *The Tender Carnivore and the Sacred Game.* New York: Charles Scribner & Sons.
———. 1959. "A Theory of the Value of Hunting." In *Transactions of the Twenty-Fourth North American Wildlife Conference.* Washington, DC: Wildlife Management Institute.
Sitwell, George. 1909. *On the Making of Gardens.* New York: Charles Scribner & Sons.
Solomon, Barbara Stauffacher. 1982. "Green Architecture: Notes on the Common Ground." *Design Quarterly,* No. 120.
Stewart, George. 1954. *American Ways of Life.* New York: Doubleday & Co., Inc.
Streatfield, David. 1970. "The Tyranny of the Garden." *Landscape Architecture* 60, January: 96–100.
Sun, Xiaoxiang. 1987. "Chinese Gardens." Paper presented at the Department of Landscape Architecture, University of Illinois, Urbana.
Treib, Marc. 1979. "Traces upon the Land." *Architectural Association Quarterly* 11, no. 4.
Tuan, Yi-Fu. 1984. *Dominance and Affection: The Making of Pets.* Englewood Cliffs, NJ: Prentice-Hall, Inc.
———. 1974. *Topophilia: A Study of Environmental Perception, Attitudes, and Values.* Englewood Cliffs, NJ: Prentice-Hall, Inc.
Ulrich, Roger. 1983. "Aesthetic and Affective Response to Natural Environment." In *Human Behavior and Environment,* vol. 6, ed. Irwin Altman and Joachim Wohlwill. New York: Plenum Press.
Warner, Sam Bass, Jr. 1987. *To Dwell Is To Garden: A History of Boston's Community Gardens.* Boston: Northeastern University Press.
Williams, Robert. 1987. "Rural Economy and the Antique in the English Landscape Garden." *Journal of Garden History* 7, no. 1: 73–96.
Wohlwill, Joachim. 1983. "The Concept of Nature: A Psychologist's View." In *Human Behavior and Environment,* vol. 6, ed, Irwin Altman and Joachim Wohlwill. New York: Plenum Press.

Luminous Technologies and Nighttime Garden Enchantments

LINNAEA TILLETT AND BRENDA J. BROWN

Sometimes, upon a trip to a new and strange place, on a spring morning, or after a drink more than usual, the filter drops away and we look at things afresh, with that sense of aliveness characteristic of fantasy. The great goal of education, for the architect and especially for the layman, should be to awaken that sense, to eliminate the emphasis on history, technology, philosophy and esthetics and cultivate instead a joyous sense of the *emotional immediacy of architecture*.

—ROBERT B. RILEY, "Architecture and the Sense of Wonder"

A garden is about nature. . . . [The garden's] power to generate fantasy, both shared and individual, lies in the fact that it is an artificial place, made from the stuff of the real. It is real enough to be convincing but not so real as to constrain the unreal. That is a powerful form of meaning, or a means to it.

—ROBERT B. RILEY, "From Sacred Grove to Disney World: The Search for Garden Meaning"

As a landscape lighting designer, I work at night, for me not a specific time but a process that begins at twilight and ends with sunrise.[1] I work in spaces open to the sky and the elements, populated by plants and trees that span ground and sky and provide habitat for a myriad of nocturnal species. I have worked in public landscapes—housing projects, campuses, parks and the exterior areas of cultural institutions—but also, as will be emphasized here, in private gardens.

Lighting gardens presents different and, in some ways, tougher, more complicated, more formal, and potentially more poetic challenges than lighting public spaces. As Riley described, gardens have the potential for experience and meaning other than, perhaps subtler and more rarefied than, public landscapes.[2] If the job of the lighting designer is to toggle between the functional and the poetic, gardens afford numerous and more delicate opportunities for the latter.

My approach to landscape lighting is shaped by three major factors: my formal training as an environmental psychologist, my exposure to ecologists and habitat conservationists, and my collaborations and interactions with landscape architects and clients. There is also the matter of my own sensibility, a thing hard to get hold of, but variously interweaving with the aforementioned influences.

Work with different astute landscape architects has sharpened my awareness of the many different aesthetic possibilities one landscape can offer. While lighting designers are the ones primarily responsible for nighttime experience, we shape the light to reflect and amplify a landscape design firm's particular sensibility. In private work, though we still likely work with other designers, we have more face-to-face interactions with clients and their individual concerns and desires come to the fore for us.

Through exposure to ecologists and conservationists I have learned how overlighting and invasive lighting can alter and destroy habitat—exhausting fish, misleading bats and birds, and disorienting turtles, as well as interfering with humans' sleep patterns. Maintaining habitat has become an ongoing concern and objective. I am thus compelled to consider less invasive lighting—solar-powered, portable, and luminescent options—as well as more conventional solutions.

Perhaps most fundamentally, my formal training as an environmental psychologist has established for me that light links us to our surroundings; it connects to our psyche, to our poetic and emotional lives. In the public realm the most obvious emotion pertaining to light is fear—fear of the dark and fear of other people and animals you might meet therein. Whether high or low income, high or low crime, our neighborhoods' usual over-riding desire regarding light is to have more of it. Political and administrative entities have related, more specific fears—fears concerning liability and lawsuits, fear of somebody tripping and falling—along with, increasingly, an awareness of light's importance in facilitating people's nighttime, outdoor, social lives.

At the same time the arts reveal a widespread sensitivity to sunlight's fragile beauties.[3]

Poets write about it. Painters paint and repaint light's qualities as they change over time; photographers engage in analogous pursuits. I have come to believe that nighttime possibilities are equally poetic, that one can bring out the more mysterious, ephemeral, wondrous, and delicate qualities of night. By working with these qualities I seek to enlarge the emotional palette. This ambition may come closest to what might be called a lighting sensibility—or at least its realm—a realm seemingly generous enough to encompass paintings' chiaroscuro,[4] memories of light's delicate dance in my parents' studio, perceptual play reminiscent of Robert Irwin's scrim experiments, and Arthur Rackman's and Edmund Dulac's illustrations of classic fairy tales and plays.

> The purpose of illuminating a garden at night is to create, as if by magic, a fairyland scene removed from reality.
>
> —GEOFFREY AND SUSAN JELLICOE ET AL., *The Oxford Companion to Gardens*

Robert Riley's 1988 essay, "From Sacred Grove to Disney World," deepens my understanding of my garden endeavors and offers hope that their fruits be meaningful. Riley clearly articulates gardens' essential paradox: they are of and about nature, but they are artifices, constructed of the stuff of the real. Crucially, he sees gardens' fantastical potential as integral to this paradox. He also recognizes, even embraces, technologies' potential role in gardens and the fantasies they might generate, though he gives little attention to the illumination or imaginative potential of nighttime gardens.

If gardens are "other worlds," they are still *other* worlds at night. They stir different perceptual modes. Eleanor Perenyi describes the heightened sensitivities to sound and scent in nighttime gardens but concludes that "the biggest change is that of proportion and texture produced by seeing things in black and white. . . . To see things in black and white is to see the basics."[5] Nighttime gardens are places of outlines, undifferentiated shapes, and reflection—without detail, without color. At night my eye is anchored by the garden's muscles and skeleton: the trees' branches against the sky, a shed's mass, a gate's outline. One sees what resists light: earthen mounds, massive rocks, tar-filled tarmac—and what catches it: schist, mica, white pebbles, an aluminum sculpture. . . . A part of me prefers the nighttime

Enchantment. Illustration by Edmund Dulac. From *Shakespeare's Comedy of the Tempest with illustrations by Edmund Dulac,* Act 5. (London: Hodder & Stoughton, 1908)

garden without my professional interventions. I would like to rely on the moon's occasional caprices, or if need be torchlights or lanterns of fireflies. On the other hand, these basic elements offer clues and cues for what I might bring to the garden's experience.

If gardens are "elsewhere,"[6] or one more representation of something else,[7] fantasy is integral even to their initial concept. One can also think of various ways in which fantasy, or *a* fantasy, has been made explicit or concrete in gardens—from the obscure tale largely entrusted to Bomarzo's sculptures, to the constructed "ruins" of Leasowes and Hagley, to the hermits hired or fabricated to populate real and constructed grottos, to the Victorian and Edwardian gardeners never to be seen by the estate owners or their guests.[8] Then too there are Lassus's theoretically grounded evocations of the unlimited that Riley describes,[9] and George Sitwell's cultivated imaginings that border on the philosophical.[10] While the life-death dialectic is integral to gardens—even high art ones—gardens' otherness apparently makes our encounters with this basic duality safer.[11] As Moore et al. observed, a garden pilgrimage will likely "circle back to where it began—like a fairy tale in which, however dragon-strewn the adventure, the young protagonist is sure to be home for tea."[12] And so, perhaps in gardens, fears of the dark can be reduced to delicious frissons and become fertile ground for a lighting designer's suggestive play.

We should acknowledge technology, not flee from it. There is no more reason for us to ignore computers and holograms, light shows and lasers, in our gardens than there was for the Victorians to ignore hydraulic and heating technologies in theirs.

—RILEY, "From Sacred Grove to Disney World: The Search for Garden Meaning"

A landscape lighting designer's work is fundamentally and entirely dependent on technology. Indeed, the history of lighting is so enmeshed with the history of its technologies that it can be difficult to disentangle noteworthy designers' styles from the technology available at a particular time. Fire, candles, fats and increasingly efficient oils, kerosene, electricity—all these technologies found their way into gardens, as they did into interior and urban environments, though not always with equal ease or success. Electricity's transformation of North American interior lighting, for example, began in the 1880s;[13] however, by 1902, Luis

Bell was noting in *The Art of Illumination* that "the lighting of outdoor spaces involves not a few unusual difficulties."[14] Outside of street lighting, world's fairs, and signage, for a long time electric lighting in exterior spaces—especially private spaces—was extremely limited. Indeed, even in the 1940s and 1950s much North American garden lighting relied on light shed from the building's interior, sconces at doorways, or pathways lit with lanterns hung from jockey figures. When lighting designers were employed, they responded to clients and architects preoccupied with views from the inside out. Indeed, in an article for *House and Garden* in 1956, after observing that "Picture windows pose problems at night too," Richard Kelly prescribed a cure for "that black mirror aspect: Use the terrace and garden as 'wall-paper' to ornament, enlarge, or even furnish the interior. To 'kill' the mirror effect, light the ground line strongly just outside the glass with lights mounted in the eaves. Selective lighting creates the perspective of the garden beyond."[15]

It was improvements in High Intensity Discharge Lighting (HID) that gave impetus to the development of the landscape lighting. First commercially available in 1932, HID was used extensively in street and tunnel lighting, but it gave off a dull, orange, unflattering glow; it turned colors to mud and faces eerie, and was thus unsuitable for highly designed gardens. Mercury vapor lights were also available commercially around this time, but many found their greenish blue light unflattering. Not until the mid-1950s were there major improvements in HID lighting, providing landscape designers with real choices for their designs. High-pressure sodium emitted a yellowish bright light—unlike the muddy orange of its low-pressure lamp cousin—and metal halide emitted a bright clear white light (unfortunately today many associate this source with sports stadium lighting). In either configuration these HID lamps produced broad beams of high-powered light with a very long life span, which meant they could be mounted at tree heights and serviced only every three or four years.

And so different styles and approaches to landscape lighting began to emerge in North America and Europe. Some designers took advantage of high-pressure sodium's vivid, clear, yellowish light. They developed styles in which it seemed rays of sunlight came down through the treetops to play on the ground and strong yellow beams were cast on building facades to create a "day for night" effect. In more tropical regions these clear yellowish beams were used to uplight palm trees or, from the top of these same palms, downlight plantings around the trees' base. Other designers relied more on the cooler colors of metal halide, adding blue gel filters and mounting units high up in trees so that the light cascaded down between the

Creating nighttime texture with light and plantings: landscape lighting by Janet Lennox Moyer. (Photograph by Mary E. Nichols, courtesy of Janet Lennox Moyer)

trees' branches for an effect akin to moonlight. These HID technologies were effective and durable. They are, in many cases, still in use. However, they were and are expensive to install, bulky and difficult to hide at ground level, and they use prodigious amounts of energy, so they are used in relatively few private gardens.

In contrast, the multi-reflector (MR-16) technology introduced in the late 1970s was relatively broadly available and made more subtle ground- and tree-mounted lighting possible. Small incandescent lamps had previously been deemed undesirable because they had a short life span and 60 percent of their energy was lost to heat; however, the new multi-reflector lamps concentrated light forward, pushed heat backward, and lasted much longer. Their fixtures were small enough to fit in your hand. One could direct light to specific landscape elements—a fountain, a tree, a flower bed—and by employing lenses, diffusers, grills, and shades, that light could be manipulated, refined, and accessorized. Sixteen- or even twelve-volts was much less dangerous to work with than the 110 volts per hour typically involved previously. And even as this technology made lighting relatively accessible to nonprofessionals, it also allowed garden lighting designers such as Janet Lennox Moyer to work at a finer grain, to treat gardens and vineyard estates as entire painterly compositions.[16]

However, in the past fifteen years light-emitting diodes (LEDs) have become dominant in landscape projects. Extremely bright and even smaller than MR-16s, they have been justly promoted for their capacity to last longer, die more gradually, use less energy, and require less robust wiring than other technologies. One watt of LED light can be as bright as a twentieth-century 30-watt incandescent bulb. LEDs do have drawbacks. They require complicated technological backups such as transformers and drivers, they can be complicated to dim, and they are sensitive to heat and flooding. They are far from panaceas. Yet because LEDs can be so tiny, because they require such miniscule amounts of power and that power can be recharged by the sun, a computer, or an electrical outlet, portable lanterns and fixtures are possible. They can serve as our contemporary "candles."[17]

Moreover, recent forms have become "intelligent," meaning a designer can "talk" to them remotely, and shuffle through a spectrum of cool light to warm, color or not, via portable hand-held devices rather than via a clunky, centralized control system. Lanterns can be carried and hung where they are needed for special entertainments and then returned to storage when the weather becomes inclement or when they might interfere with wildlife. Large events such as weddings and concerts can be lit without large generator trucks. In places where power is unavailable solar lights can be installed without digging up and running new electrical wiring through planted areas. One can specify portable lighting and design its shape, weight, and the stake on which it will hang. One can break from the model in which an electrician is necessary and special approvals must be obtained from government bodies. The lighting designer can install the lights and can also change them, as well as their programs or locations when seasons turn or the client wants garden additions or subtractions.

Today the nighttime lighting designer's job is not to pick fixtures from a catalog but to communicate with and give shape to visible energy. One selects how each part of the visible spectrum will work—its intensity, its brightness, its color, its direction. One chooses light sources—candlelight, incandescent, HID, LED, solar—to determine how light will behave; and one decides when different lights will be on and off. One controls how the energy is to be brought to life, be it an alternating current or a simple solar panel. One shapes visible energy so as to create or support a landscape's given or desired emotional texture.

A garden is about fantasy. A garden can be not only a fantasy realized, but a fantasy provided . . . the garden can be seen as not a concretization of fantasy, but as a spur to fantasy.

—RILEY, "From Sacred Grove to Disney World: The Search for Garden Meaning"

The technological revolution shaping nighttime landscape lighting in the thirty-plus years since Riley's essay is undeniable and it continues apace. It is nearly impossible to keep up with new lighting technologies and their possibilities. These developments have implications for human experience and for the experience of other living things.

If the computers and light shows—if not the holograms and lasers—to which Riley referred have become commonplace, we have also discovered reasons not to employ them. The Tribute of Light, activated each September 11 in lower Manhattan, has become famous for the 160,000 birds whose migrations it disrupts as well as for its evocative power. While steps have been taken to lessen harm by periodically turning off these lights,[18] the project also stands as a dramatic example of how city lights can confuse and exhaust birds and increase their collisions with skyscrapers' glass walls. Although one garden's lighting is unlikely to instigate such a large-scale disaster, consider the cumulative effects of many garden lighting practices: birds and insects smashing into lights, moth populations booming, and fish habits and habitats deranged. Fostering biodiversity means living with some darkness; this probably means adjusting our expectations for how landscapes are lit. The technologies Riley cited tended to engender fascinating on-the-ground spectacles that outshone the natural sky's demure points of light; newer technologies make other sorts of effects and affects possible. While many other creatures may benefit from them, they also have implications for human garden experience and meaning, particularly in regard to fantasy, enchantment, and the play of imagination.

(Party) lighting among the crepe myrtles. (Lighting rendering by Tillett Lighting Design)

In lighting almost any garden one works to fulfill the client's needs, needs that are functional and needs or desires that are more emotional and usually much less explicit. Function of course is very important and it is not necessarily simple to use light instrumentally to support circulation and programmatic activities and guard habitat at the same time. But one can take lighting to the next step, to use light to delight, intrigue, and evoke, to create or help create a garden that means something to someone. One might use light to create a place of safety for the client's individual fantasies and perhaps a place that tickles others' imaginations. Light may be used to give different parts of the garden different qualities: a part that has a poetry, a part in which one senses things growing, a part that captures the color and a memory of the flowers that once grew there. In the garden's safe containment, light can maintain or enhance nighttime's mysteries and gently ensnare visitors in enchantments of interacting light and darkness.

While many artists and writers have evoked light's entrancing and poetic qualities, within the lighting profession the vocabulary to describe its evocative effects remains underdeveloped. Perhaps no one has done it better than Richard Kelly, so-called "Father of

Architectural Lighting." Kelly's lighting for Philip Johnson's "glass house" epitomizes the modernist preoccupation with the view from the inside out rather than experience within the garden itself. Nevertheless, Kelly developed a sensitive and expressive language for describing light's subtle aspects and experience, a language encompassing both the subjective and the objective.[19] He understood the power of light to affect our awareness at conscious and unconscious levels; after his visit to one site, for instance, he reflected in a memo that lighting's distribution there needed to be done with great care, its intensities subtly controlled to result in an effect "almost below the level of conscious awareness."[20] If light is used deliberately and sparingly its effects—and affects—can be greater. It can work on the edge of people's consciousness. If you think about what you really want people to feel, you can use less light. New technologies make the realization of such intentions more likely.

While the entire array of twentieth- and twenty-first-century lighting technologies are available now, earlier ones exert continuing attractions: white-painted stones to reflect light, the whitish plants Vita Sackville-West and Harold Nicolson used in Sissinghurst's White Garden, or a candle-held lantern hung in a tree. However, most people have been unwilling lately to put torches or candle-light lanterns out into their gardens except very briefly. Yet, perhaps ironically, after years of enthusiastic and sometimes hyperbolic lighting, recent code restrictions and environmental regulations concerning night sky and light trespass as well as a growing awareness of landscape habitats are requiring smaller gestures. Bell presciently maintained that "in lighting generally . . . a little tact and skill can affect an economy far greater than all the material improvements of the last 20 years."[21] Supporting a nighttime landscape's nature and emotional texture today requires "tact," a sensitivity to the project's human and nonhuman dimensions and "skill" in understanding how to situate and operate new lighting technologies. While the new regulations and constraints are not agreeable for those who seek drama and theatrical effects they suit me. Perhaps my sensibility thrives within them. Today's more nimble equipment allows one to better relate to light as energy rather than as something presented as a package. Smaller and increasingly sophisticated, it makes subtle affects and effects more plausible. We can now think of equipment as a sort of cauldron of energy, light that can be directed but that is movable too.

Riley describes gardens as artifices as well as particular articulated expressions of humans' relationships to nature. Artificial lighting is not natural; but it can play a vital role both in gardens' artifices and garden's constructions of nature. Somewhat paradoxically, handled carefully in the garden, it can make nature's nighttime wonder and mysteries more accessible by enticing people to go where they might not otherwise go and making it comfortable for them to linger there. Tiny lights flickering like candles may mark a roadway's edge; arriving guests may be given a small light to carry into the landscape to mark and illuminate their own space. The illusions and fantasies light technologies make possible may be as explicit as an off-season evocation of a lavender field's purple, or as subtle as a now-you-see-it now-you-don't suggestion of fairies dancing atop a pergola. But they will never be in the same league as fireflies and the moon.

NOTES

1. While this essay is the work of two authors, the "I" is Linnaea Tillett.

2. Robert B. Riley, "From Sacred Grove to Disney World: The Search for Garden Meaning," *Landscape Journal* 7, no. 2 (1988): 136.

3. There are also of course examples of painters who have dealt with night light, for example, Robert Albert Blakelock and Albert Pinkham Ryder in the United States.

4. "I propose, therefore, to transfer the aesthetic values of the painter's chiaroscuro to the realm of the psyche's aesthetic values" (Gaston Bachelard, *The Flame of a Candle* [Dallas: Dallas Institute Publications, 1961]).

5. Eleanor Perenyi, *Green Thoughts: A Writer in the Garden* (New York: Vintage Books, 1981), 142.

6. Isabelle Auricoste, "Leisure Parks in Europe: Entertainment and Escapism," in *The Architecture of Western Gardens* (Cambridge: MIT Press, 1992), 483; Brenda J. Brown, "Landscapes of Theme Park Rides: Media, Modes, Messages," in *Theme Park Landscapes: Antecedents and Variations*, ed. Terence Young and Robert Riley (Washington, DC: Dumbarton Oaks Research Library and Collection, 2002), 235.

7. Robert Harbison, *Eccentric Spaces* (Boston: David R. Godine, 1988).

8. Ronald Blythe, *Akenfield: Portrait of an English Village* (New York: Delta Books, 1969).

9. Riley, "From Sacred Grove to Disney World."

10. George Sitwell, *On the Making of Gardens* (New York: Charles Scribner & Sons, 1909).

11. Yi Fu Tuan, *A Study of Environmental Perception, Attitudes and Values* (Englewood Cliffs, NJ: Prentice Hall Inc., 1974).

12. Charles W. Moore, William J. Mitchell, and William Turnbull Jr., *The Poetics of Gardens* (Cambridge: MIT Press, 1993), 118.

13. With the lighting of J. P. Morgan's mansion in 1882.

14. Luis Bell, *The Art of Illumination* (New York: McGraw Publishing Co., 1902), 244.

15. Richard Kelly, "Garden Lighting," *House and Garden*, July 1956, 10.

16. Moyer's *The Landscape Lighting Book* is a seminal work on landscape lighting design: Janet Lennox Moyer, *The Landscape Lighting Book* (Hoboken, NJ: John Wiley & Sons, 2005).

17. Luis Bell presciently remarked: "If electric light had been in use for centuries" and the candle been just invented it would be hailed as one of the greatest blessings of the century, on the grounds that it is absolutely self-contained, always ready for use, and perfectly mobile." Electricity has been in use for well over a century and these tiny LEDS have now given us the possibility of mobile, self-contained, ready for use, contemporary candles.

18. Anne Barnard, "The 9/11 Tribute Lights Are Endangering 160,000 Birds a Year," *New York Times*, September 9, 2019, retrieved October 15, 2020, https://www.nytimes.com/2019/09/09/nyregion/911-tribute-birds.html.

19. Kelly defined three lighting elements, and these definitions have their own poetry: the focal glow (the campfire of all time), ambience (a snowy morning in open country), and the play of brilliance ("a cache of diamonds in an open cave").

20. These reflections came in a memo Kelly wrote after a site visit to the Lodge at the Rockefeller estate. Reference was made in an unpublished manuscript by Tillett and Gardner, but the original memo can no longer be located.

21. Bell, *The Art of Illumination*, 209.

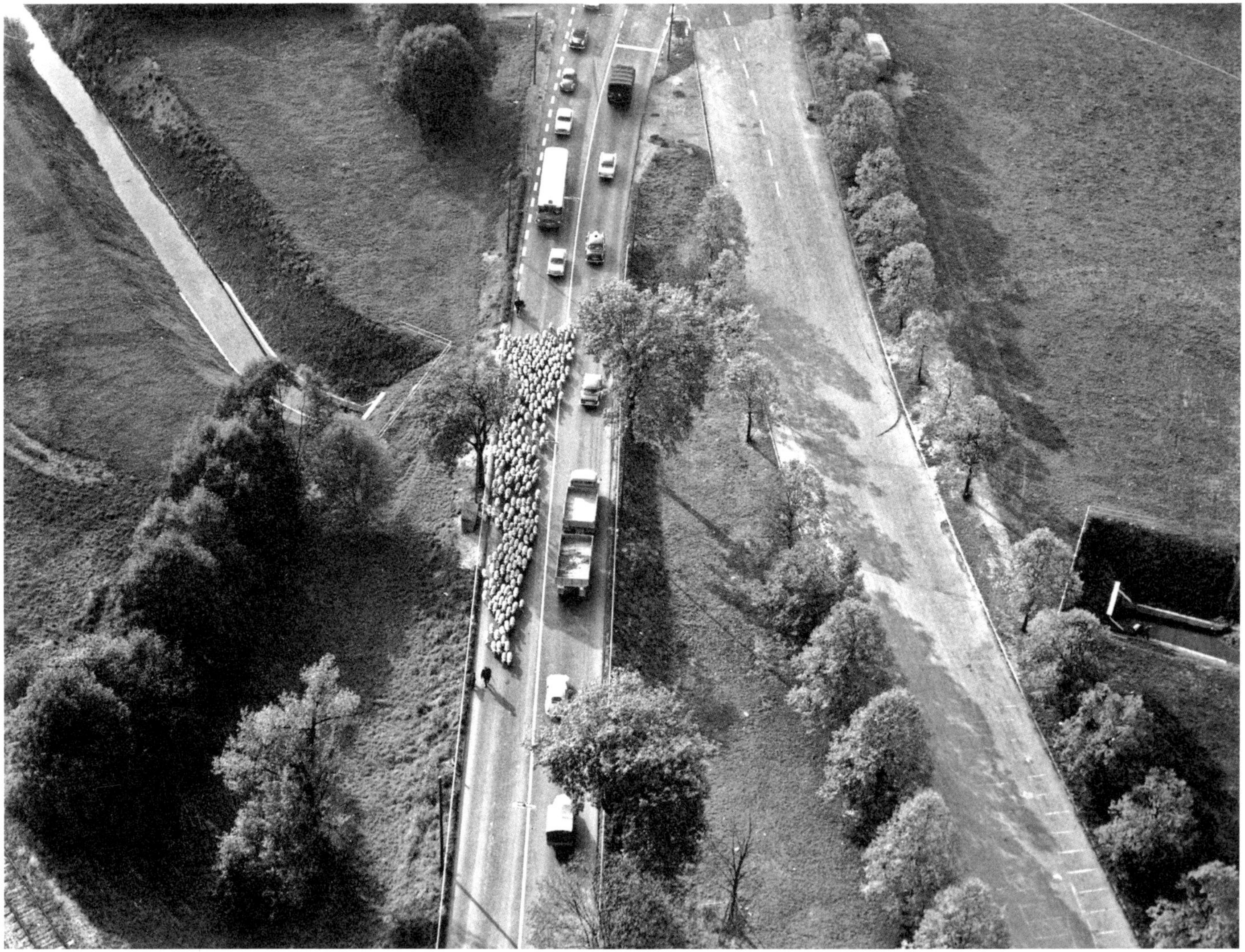

Flock of sheep on highway.
(From Hans Reich, *The World from Above,*
1966, © www.luftbild-bertram.de)

V. Landscapes and the Road

1968

Auto Territoriality

ROBERT B. RILEY

Considering the design profession's current preoccupation with the psychology of spatial experience, it is strange how little thought has been given to the behaviorist implications of the automobile. The planners' dislike for accelerating American dependence on the automobile is a major determinant of much contemporary urban design. Unrestricted use of the private car docs have its disadvantages. It causes pollution. It is a markedly inefficient method of peak load transportation to and from the crowded centers of our cities. It demands heavy expenditures in the public sector and, proportionately, perhaps even higher private expenditures for the two-car commuter family. It magnifies the already difficult problem of preserving wilderness areas. Most seriously of all, it is producing a new city shape which the planning profession, educated in traditional concepts of urban physical structure, has so far been unable to cope with.

But the anti-automobile bias of the professionals seems to be based upon more than such rational objections. Putting aside supercilious descriptions of the American's car *as* a lover-image or a totem, the fact remains that the automobile does have a meaning that for many people reaches far beyond its function—an emotional attachment represented by the weekend rituals of washing or tinkering. It might be just this attachment, as much as its functional limitations, that explains the low esteem in which the automobile is held by planners. Somehow, strolling the streets, sipping wine at a sidewalk cafe, playing chess in the park, sampling the art galleries or gathering for darts at a pub are all considered worthwhile pastimes for the urban dweller, while repairing or grooming cars is somehow distasteful. This is a distinction that smacks at bottom of class-consciousness or snobbery. It is a reflection of the profession's unspoken vision of the city as primarily a stimulating setting for the urbane connoisseur.

But far more dangerous than the planners' emotional aversion to the automobile is their total failure to understand the reasons for its popularity. The kind of pseudo-Freudian analysis popularized by John Keats and others is more useful in justifying existing prejudices than in furthering real understanding. And yet in the collection of contemporary planning concepts there exists a powerful and obvious tool for the understanding of the automobile's role—the concept of territoriality.

There is, to be sure, more than a bit of gobbledygook and silliness and plain misunderstanding in the designers' current preoccupation with some of the newer concepts of the psychology of spatial perception. But work of Hall and others has shown, plainly and surely, that men move through. build around them, and carry about with them, certain structured volumes of psychologically differentiated space. There is doubt as to how these spatial volumes are best understood, and doubt as to their exact meaning, but there should be no doubt as to their existence or their—at least sometimes—importance. Nor should there be doubt that constriction or expansion of these distinct spaces, or conflicts between them, or ambiguities as to their boundaries, can cause uncertainty and stress.

This idea of man as living within a range of simultaneous spatial domains graded from the intimately personal to the plainly public should explain, more satisfactorily than any of the many familiar psychiatric or sociological clichés, the attraction of the automobile. Put simply, the automobile allows one to travel almost at will anywhere in the public domain while remaining in a completely private world unequivocally defined by physical boundaries. The maintenance, defense or even definition of this intensely personal space no longer needs to be achieved by psychological adaptation or cultural understanding or ritual. It is marked off structurally—with clarity and solidity. Perhaps Americans are more concerned than most peoples with the definition of personal space—though Hall claims that the Germans are even more jealous of its boundaries. It might be that the rapid and continuing changes in the American social scene over the last fifty years have produced a general uncertainty and unease that places more importance and value upon the protection and clear definition of the private personal realm. While the traditional Utopian visions have been built around a communal structure, modern Americans are attempting to build very personal or at least familial Utopias—Utopias structured around detached houses, television and automobiles. There is at any rate a basic distinction between the automobile and other methods of transportation that far transcends convenience and economy, a distinction that must be understood. It is the distinction between *public* and *private* transportation, not in the sense of financing or titular ownership or even trip scheduling, but in the sense of the *personal perception of space patterns.*

This ability to move through public space without suffering impingements upon, or readjustment of, one's own personal space could explain much more than the commuter's attachment to his private automobile. It might partly explain the phenomenal success of the car rental companies, for a rental car allows one to travel a strange and foreign and often confusing public world in a kind of instant privacy and encapsulated security. It might partly explain, as much as laziness, the propensity for taking automobiles into the "wilderness" areas—areas that by definition lie at the opposite extreme from personalized space. Surely it explains much of the success of the motel business, for a motel is more than just convenient—it enables one to move between the personal spaces of car and bedroom without traveling through the more public and often spatially ambiguous realms of lobby, elevator and corridor associated with the traditional hotel.

It must be understood that the automobile is not just a more or less efficient competitor of public transit, nor must the emotional importance of the automobile be contemptuously dismissed as some psychological aberration. It might be that in certain parts of our older cities the side effects of the private automobile are so space and time consuming, so physically wasteful or even unhealthy as to require restrictions on its use. It might be too, that the only way of increasing the acceptance of "public" transportation is to incorporate into it as much of the spatial quality of the automobile as is possible. But the transportation problems of our urban regions are not going to be solved until we admit to an understanding of the very special spatial features of the automobile, for it is likely that the appeal of the automobile is not just a silly habit easily gotten rid of, but a means of fulfilling deeply rooted concepts of human territoriality.

The Overlanders

Technology and Culture on the Pan-American Highway

ROSA E. FICEK

They drive for months at a time, seeking refuge from the soul-crushing grind of late capitalist civilization. The marvelous vehicle-homes that take them into unknown lands are equipped to withstand the harshest off-road conditions, yet also offer interior space inventively fitted for all aspects of living. Cocooned in these mobile shelters they embark on lengthy expeditions across the continents of North and South America, leaving a digital trail in blogs and social media that gives clues about what they search for, and what they find.

United States motorists who take long-distance road trips on the Pan-American Highway—overlanders, as they call themselves—describe an extraordinary, life-changing experience.[1] If, as Robert Riley suggests, we are only beginning to craft analytical approaches that can help us more fully comprehend the landscape as both an affective force and a site of human interaction, the travel writings of overlanders offer a way to understand how both the social life of the landscape and the mood it creates in people together help shape particular landscape experiences.[2] Over fifty years ago Riley observed how automobile travel, in particular, is emotionally appealing to US drivers because it allows them to preserve their personal domestic space even as they experience public spaces such as highways and natural areas.[3] The travel writing that overlanders produce demonstrates the tensions between the personal and the public domains, the car and the road, inherent to this form of engagement with the landscape. A reading of overlander travel writing about Latin America through Riley's insights, together with critical attention to the history of US projects in the region, further reveals how these tensions are shaped by the unequal relationship between the nations that the Pan-American Highway connects.

Overlanders chase awe-inspiring wilderness beyond the borders of the United States in an effort to connect with themselves and with the world. This essay analyzes overlander travel writing in relation to the history of US expansion in Latin America, expansion that makes it possible to experience wilderness through the Pan-American Highway. Special attention is paid to how overlanders negotiate differences of race and class on the road. While overlanders do not name the power relations that give them access to faraway places, their alignment with expansionist projects is evident in the ways overlanders interact with the people they encounter who are different from them, and in their use of technology to "conquer" the road—to assert US agency and control over places that belong to other people. And yet, attention to the affective impact of the road on the travelers also reveals the precariousness of their high-tech wilderness experience. The road's bumps and turns bring overlanders into contact with the local people who are written out of their travel narratives. When considered through the lens of US empire, these encounters generate questions about the transformative power of overlander journeys and the limitations that personal vehicles present when engaging with landscapes of the public domain.

The Overlanders

Overlanders on the Pan-American Highway seek escape. Their writings hint at former lives confined in offices and meetings, corporate careers, material comforts, and a meaningless existence. Then, one day, it happens: it is time for a change. The Pan-American Highway beckons. This journey of personal transformation calls for undoing ties, as one overlander puts it, a shedding of material possessions and distancing from home and family. They quit jobs, sell their things, visit loved ones one last time. As they dismantle their old lives into bits and pieces, they assemble with care a new home on wheels. Overlanders lovingly modify trucks and vans in anticipation of difficult road conditions and off-road driving. Many look forward to spending time in nature and equip their vehicles with facilities for sleeping, bathing, and cooking. They make sure to remind their readers—potential travelers—not to forget essential camping, photography and communications gear.

Altogether, these items and modifications allow the vehicle to withstand arduous travel and the travelers to withstand overnighting in nature. In other words, they are able to drive scenic roads, camp, and hike in lands unknown to them. In fact, many take pains to distance themselves physically and categorically from tourists, preferring instead to identify as explorers.

The technology of vehicles and gear also allows overlanders to make their own route. While driving the Pan-American Highway may be their stated aim, none actually stay on the highway's route. Instead, their travels are guided by the pursuit of life free from capitalist white collar drudgery. As they chase wild and spectacular destinations, they photograph, film, and write about their experiences in self-published blogs that are part guidebook and part first-person travel account. In these blogs and other social media pages, overlanders describe sites both on and off the beaten track. They evaluate hotels, campsites, and restaurants; they narrate their experiences on the road with their vehicles and getting to the sites. The blogs transport readers to secluded coastlines, pyramids, hot springs, museums, and sidewalk cafes. Photos are prominent, and depict various configurations of landscape, vehicle, and road. Overlanders photograph the landscape from the perspective of the road, and the road from the perspective of the landscape. Readers also find many images of the vehicles themselves, as part of the roadscape and other times just sitting in a parking lot. Sometimes the blogs include photos of wildlife, or snapshots of the overlanders socializing with other overlanders. But most of the visual content is dominated by spectacular photos of beaches, mountains, ruins, and cities.

The Expansionist Underpinnings of US Driving Culture

Why drive the length of the Pan-American Highway? When one blogger posed this question to the route's postindustrial explorers, responses revolved around the fact that, especially for those in the United States, the road is just *there*. Given its closeness and accessibility, why *not*? More than the places that the highway stitches together, their responses suggest, what matters is the experience of the road itself. And central to this experience is the scale of the road trip. The point is to live on the road for an extended period of time, and to travel the entirety of the Western hemisphere's two continents. While these road trips offer escape from economic lives very specific to the early twenty-first century, the practice of jumping into a vehicle and driving into unknown lands has precedents in the history of US expansion.

The desire and the anticipated experience of driving the Pan-American Highway are made possible by decades of US cultural, economic, technological, and political interventions in the region.

In the United States, leisure driving culture is linked to the enjoyment of nature in ways that legitimize colonial violence. In contrast to Indigenous ways of knowing that consider humans an extension of the land and the kin of other living beings, Euro-American ways of knowing position humans outside of nature.[4] In a settler colonial society such as the United States, wild, natural places are the result of an ongoing process of Euro-American occupation that eliminates, obscures, or erases Indigenous people from the land.[5] Leisure automobile driving, within this context, reinforces the structures of US expansion by allowing people to access and consume nature without acknowledging the people who belong to the land, and to whom the land belongs. Their access to that land is fundamentally predicated upon native erasure.

Motorists have admired nature from their vehicles from the very early days of automobiling, when country tours became a popular pastime. Like contemporary overlanders, early twentieth-century motorists found it necessary to bring along gear to withstand rough road conditions. Along with goggles, protective clothing, and mechanical tools, many early drivers also brought along camping gear. As the historian of driving Warren Belasco shows, in the 1920s autocamping became popular among motorists who preferred to camp on the side of the road or in natural areas rather than stay in hotels, which at the time were linked to the railway system. Autocampers rejected the timetables and routes of railway travel and the stiff formalities of resort hotels. For them, the road offered freedom from the routines of industrial life through contact with nature. It also offered freedom from the schedules of commercial tourism. Over a short period of time, autocamper tents set up on the side of the road—or on people's lands, with and without permission—gave way to house trailers and municipal autocamps.[6] Meanwhile, changes in auto design, especially the transition to enclosed vehicles, made auto travel more comfortable and fed people's desires to enjoy natural scenery, leading to the creation of National Park roads in the 1930s.[7] Family tourism by automobile had become a means to escape into wild landscapes that provided respite from modern life.

These expressions of freedom, however, are not innocent of the violence inherent to the American landscape. The history of US expansion often focuses on westward explorers and pioneers but does not as often dwell on how settlers displaced and erased Native Americans

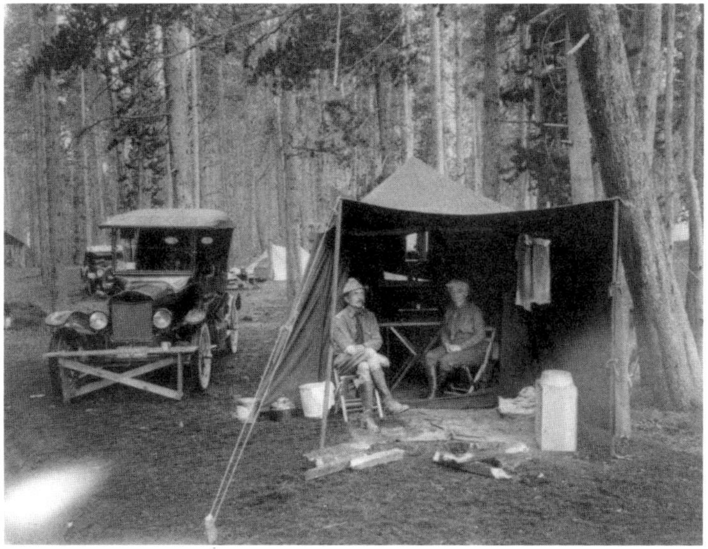

left The first auto to reach the Grand Canyon, part of a "pioneer automobile party" demonstrating the feasibility of the route, 1902. (Library of Congress)

right Autocamping in Yellowstone National Park, ca. 1923. (Library of Congress)

from the landscapes. In the myth of the frontier crafted around this past, immigrants became American by surviving the hardships of the wilderness and transforming this wilderness into a domesticated landscape. This concept of wilderness celebrates nature as a source of spiritual renewal and personal transformation. By claiming a connection to the landscape, settlers became self-sufficient individuals—the new natives—and justified the occupation of Native American lands.[8] Leisure motoring in the US often reenacts the frontier myth. Road trips to national parks and other wild areas offer a way to obtain spiritual renewal through immersion in nature despite, and in fact because of, the violent means through which this land became accessible to the United States in the first place.

The territorial control that this form of driving enacts takes on a distinctly continental scale. Horatio Jackson's 1903 trek from San Francisco to New York was the first of many transcontinental road trips across North America to promote vehicles and routes, and to prove that the journey was possible—even if the proper road infrastructure for motor vehicles did not yet exist. These grand feats of human courage and automotive technology were not limited to the United States. Motorists staged races and competitions across the globe during the early decades of the twentieth century. The 1907 Peking to Paris race and the 1908 New York to Paris race (via Russia) crossed east and central Asia, while in the 1920s the England to Australia expedition crossed the Asian continent southward.[9] In 1920s Africa, multiple expeditions and races from Cape Town to Cairo crossed the continent in a South-North direction, while other routes crossed East-West.[10] In Australia, men and women raced around the continent in the 1920s and again in the 1950s.[11]

Despite taking place in very different locations, all of these transcontinental road trips have in common an asserted imperial control over vast landscapes. A connection between motoring and competition between nations was forged in Europe in the early days of automobilism, when races and other motorized sports prepared European societies for World War I by familiarizing mass spectators with the danger of driving and the violence of crashes, and by training men for wartime service with motor vehicles.[12] Long-distance rallies and expeditions reproduced these dynamics on a broader territorial scale, allowing European states to symbolically reassert possession and control over colonies as well as internal frontiers in ways celebrating motorists' assumed technological and moral superiority over the people whose lands they traveled through.[13] In Australia, motorists borrowed the term "overlanding" from exploration narratives to describe treks around the continent. Such sporting events imagined Indigenous terrain as empty land that lacked technology and legitimized settler control over the landscape by showcasing their assumed superiority.[14]

When US overlanders say that they drive the Pan-American Highway because it is just *there*, this claim of *thereness* needs to be understood in relation to histories specific to the formation of the United States as a settler colonial and imperial state. To even *think* of driving across North and South America is possible because historically US automobile culture has depended on access to vast spaces made available for white settler occupation through violent means.

The Pan-American Highway and the United States

The idea to build a modern highway linking North and South America was proposed precisely during the 1920s frenzy of transcontinental tours, and just as wild areas in the United States were increasingly becoming available for leisure auto-based consumption. Politically, US policies toward Latin America were shifting. The Pan-American Highway, its US motor-

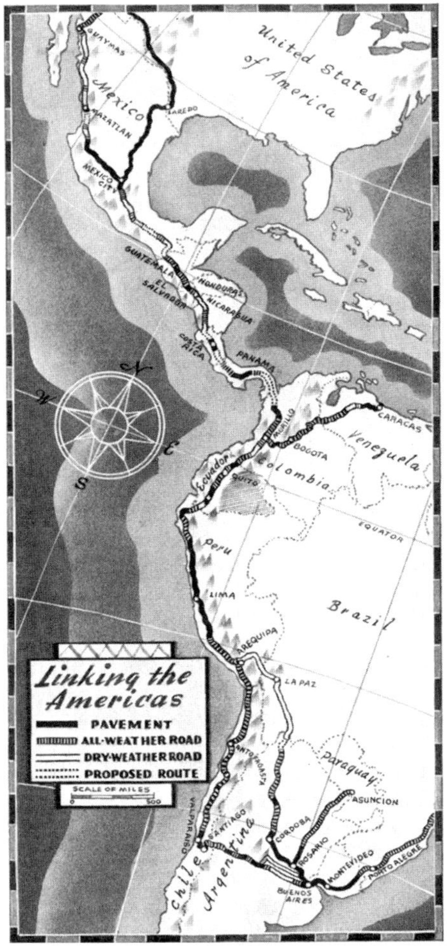

Map of the Pan-American Highway in a 1941 magazine article about travel to Latin America. While tourists tend to focus on the Latin American portion, the Pan-American Highway system includes major highways in Latin American countries, the United States interstate highways, and the Alaska Highway.

ists, and the pan-Americanist discourse that came along with promoting the route to travelers extended US expansionism, turning attention south after there were no more lands to conquer in the West.

Toward the end of the nineteenth century the United States underwent a series of economic recessions that reflected a crisis in industrial overproduction. One proposed solution was to seek out new markets in Latin America—the survival of this particular formation of liberal capitalism depended on it.[15] At the same time, many investors and policymakers considered the British Empire and its domination of global commerce a model for increasing national wealth. Expanding commerce with Latin America, they argued, would help the US economy and "civilize" Latin Americans by transforming their practices of work and consumption. The expansion of US influence in Latin America did not aim to assimilate other nations; it was justified, rather, as a moral responsibility toward supposedly inferior people who were considered so different racially and culturally that it would be impossible and undesirable to turn them into (US) Americans.[16] This project worked to control the region through commerce rather than territorial annexation.

This discourse of pan-Americanism helped generate the idea of a Pan-American Railway, which in the 1920s regenerated as the Pan-American Highway. As a political project, Pan-Americanism invested great confidence in the transformative power of technology and obsessed over the production of scientific knowledge through statistics, archaeological studies, maps, travel guides, and informative bulletins that altogether had the effect of integrating Latin America into the sphere of US capital accumulation and knowledge production.[17] Representations of Latin America made by the United States during this time positioned the region as a continent full of economic opportunities defined by a contrast between ancient cultures, once full of splendor but now in decline, and a modern culture evident in capital cities.[18] Pan-Americanism's combination of economic expansion, racism, and faith in the modernizing power of technology shaped how the Pan-American Highway was conceptualized and built, and now shapes the experiences of motorists who drive its length.

Pan-Americanism took on additional nuances in the 1930s and 1940s, just as governments across the region undertook planning and construction projects for their sections of the Pan-American Highway.[19] In the United States, the Good Neighbor policy (1928–47) represented a change in policy with respect to Latin America. Whereas before US relations with Latin America had been characterized by political and military interventions in the

The scenic beauty of the Pan-American Highway in Mexico, depicted in a 1955 travel guide.

Caribbean and Central America, the new policy abandoned displays of military force in favor of multilateral negotiations.[20] The idea of an international community comprising the nations of the Western hemisphere worked to unify a vast and heterogeneous space under the economic and political influence of the United States. Policymakers, together with the producers and consumers of US travel literature, constructed the idea of an interdependent hemispheric "neighborhood" by drawing on the idealized model of small residential communities in small-town United States made up of white, middle-class, Christian, single-family homes.[21] The Good Neighbor policy promoted a sense of responsibility toward people in other countries and sought to create in the region a closed system where the United States could exert control in symbolic and material ways.[22]

Within this context, touring Latin America by car was at once a neighborly practice and a way to enact imperial fantasies. To promote motor tours, the Pan American Union published guides with information on routes, road conditions, lodging options, and local attractions.[23] It also publicized transcontinental journeys, such as the expeditions led by Herbert Lanks and Sullivan Richardson between 1939 and 1942, which aimed to demonstrate to US audiences that travel by automobile on the Pan-American Highway was not only possible but also safe and maybe even pleasurable.[24]

Technology, Freedom, and Conquest on the Road

When overlanders cross over into Latin America today, they bring their driving culture with them. Just as the autocampers and transcontinental motorists before them, for overlanders, the experience is more important than the road itself. This experience of wild landscapes and vast distances is shaped in fundamental ways by auto and travel technology.

The size of the vehicle is an important consideration. Because they bring their home with them, overlanders need ample space inside. But if the vehicle is too large, it may not fit through certain streets in urban areas—as some overlanders have found and written about. Trailers may offer many comforts but smaller camper vans are more agile and easier to station in parking lots. At their best, overlander vehicles can manage a wide range of roads and road conditions, and more importantly, allow overlanders to access wild places, not only because of how the vehicles are constructed, but also because of the living facilities and gear they contain inside. Equipped with GPS technology, overlanders need not depend on print maps, or on locals to ask for directions. With the push of a few buttons, overlanders can travel from site to site avoiding local interactions that would transfer knowledge of the landscape from locals to the travelers. Rather, they rely on corporate satellite technology that has no connection to the sites they visit other than surveillance from afar. The vehicle's technology is essential for overlanders to make their own route as they please, free from tourist traps and bus schedules, free from the line on the map that marks the Pan-American Highway. The Panamericana is more of a suggested route, anyway.

If overlander technology creates the route, it also creates the wilderness the travelers enjoy. The landscapes through which overlanders travel have been shaped, are shaped, and will be shaped by local uses and meanings. At the same time, the technology that overlanders bring with them opens up the landscape to new uses and meanings. On the Pan-American Highway, the need for camping gear arises because the road, and the sites that overlanders seek, are for the most part not equipped for tourists. Camping gear allows them to occupy landscapes they do not know. Beds, kitchen equipment, cabinets, water storage, and so-

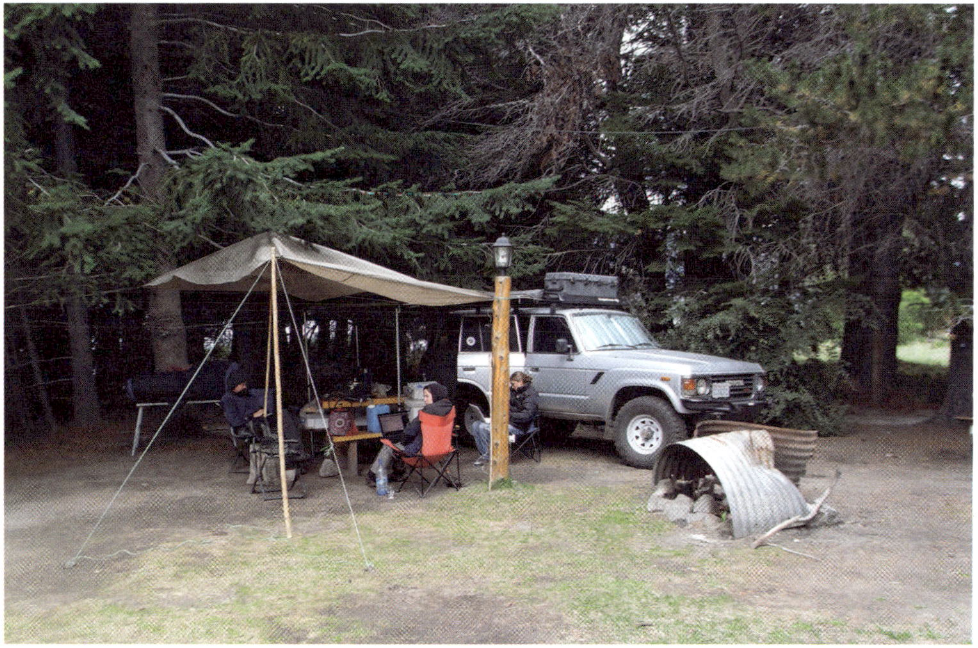

A campsite in southern Argentina. (Courtesy of Liferemotely.com)

lar panels and other power systems allow overlanders to settle in and make themselves at home. But the deserted beach where overlanders camp may also be frequented by fishermen and be an important part of local livelihoods. And the lush nature overlanders may observe on their hikes may actually be someone's farm. The wilderness experience is created through their technologies and gear. Overlanders do not go into the wild. They bring the wild with them.

A look at how overlanders represent these landscapes in their writings and images can shed light on how technology produces this wilderness experience, along with its inherent violence. Through photographs, websites, and social media, overlanders stage a wild landscape erased of local people. The sites they enjoy are the result of long histories of human engagement with nature, even if those histories are not evident to the traveler from faraway lands. Indeed, US representations of the Pan-American Highway center US-Americans in majestic

Spectacular wilderness on the road in southern Argentina. (Courtesy of Liferemotely.com)

landscapes, erasing visually and textually the people who live and work in these places. When locals do appear in these representations, they are often in service roles: the street food vendor who sells local delights to the travelers, the friendly boy who brings overlanders resting in hammocks a cold beverage. Landscapes shaped by local labor for local purposes become landscapes for leisure and consumption for US travelers. The freedom of the road is based on the erasure of people who were there before, and who continue to occupy the space.

Highway overlanders, their journey of personal transformation—into entrepreneurs—may be a way to reinvent themselves as new capitalist subjects.

Overlanding Fails

For the United States in the mid-twentieth century, the Pan-American Highway represented a utopic vision of the Western hemisphere as a community of "good" neighbors. Through this lens, Latin America was an exotic destination where northern travelers could marvel at the remains of non-Western civilizations *and* an active frontier of US-style modernity. Travel writing and other representations of the Pan-American Highway during this time emphasized the inevitability of progress. For example, Herbert Lanks, who drove lengths of the Pan-American Highway in the late 1930s and early 1940s, juxtaposed photographs of modern schools to ancient ruins while at the same time inviting potential motorists to imagine themselves as conquistadors.[27] Armed with modern technology, conqueror-motorists are agents of change who help bring about landscapes of the future. Motoring guides to the Pan-American Highway published by the Pan American Union and other travel writers in the 1960s give us a further glimpse into this future.[28] By describing tourist attractions, lodging and recreational facilities, and details about road conditions, the guidebooks promise a highway system and tourist infrastructure that, if not yet entirely modern, *will* be modern in the foreseeable future. The Pan-American Highway is described, for example, through its pavement—the Pan American Union's 1969 guidebook to motoring in Latin America lists kilometers in asphalt, kilometers in dirt road, kilometers yet-to-be paved but that are nonetheless listed as part of the highway system and drawn on the map with the certainty that the highway system—and the pan-American dream—will be a reality.[29]

After the 1960s, travel literature on the Pan-American Highway diminished until the early twenty-first century and the possibilities for self-publishing that the Internet and digital technology presented. The proliferation of Pan-American Highway travel blogs after 2008 suggest a connection between this new wave of travel writing and the 2008 financial crisis. Driving Pan-America in the aftermath of economic and political collapse—not only in the United States, but across Latin America—when dreams of hemispheric progress and development have been abandoned, what might contemporary overlanders tell us about the pan-American project as it has been defined by the United States?

Rather than as conquerors and agents of displacement, Pan-American overlanders position themselves as refugees, people displaced, in search of connection to nature, to themselves, to the world. And yet, it is the very US American idea of nature, and especially wilderness, that allows a road trip or a hike to seem like an innocent and even alluring way to spend time. Why not drive the Pan-American Highway? It's right *there*.

In true pan-American fashion, overlanders use the latest technology to conquer the landscape, making the highway adapt to their whims rather than follow the line on the map. However, driving from destination to destination as they please creates and maintains an insulated network of travelers, knowledge, and resources about the landscape. This technology, and the community of overlanders it sustains, prevents them from noticing how they actually fit into these landscapes. Their marvelous vehicles offer shelter from the elements, but also shelter them from interactions with other people. Armored with vehicle and GPS, overlanders manage to avoid uncontrolled encounters with the landscape and its people—especially encounters that may require them to shift from their position of conquistador to a different social role.

Consider, for example, what overlanders write about Indigenous people barricading roads. Roadblocks are part of Pan-American Highway driving lore to the extent that overlanders anticipate running into them and often make a point of reporting their experiences. A common strategy is to drive around by going off road or taking alternate routes—an alternative not available to just any vehicle. One blog describes a carefully timed maneuver to cross a roadblock at the exact moment the road blockers lowered the rope to let a truck drive past in the opposite direction.[30] At some of these roadblocks, overlanders write, locals demand money. In other cases, readers will never know what motivates the roadblocks because avoiding contact is an important feature of overlanders' technological armor. It can shield and remove them from potentially uncomfortable encounters in which their wealth and privilege may be the focal point of conversation—and negotiation.

The dream of a highway that promotes understanding and goodwill among the nations of the Western hemisphere has failed, to put it mildly. And yet the highway, in its variable surfaces and routes, provokes encounters with locals, pulling overlanders into the landscape as it is used and understood by the people who live there. When GPS systems fail, the road can lead them through slums and other scenes, such as roadblocks, that they would rather avoid but that nevertheless present opportunities to engage with people in ways that unsettle

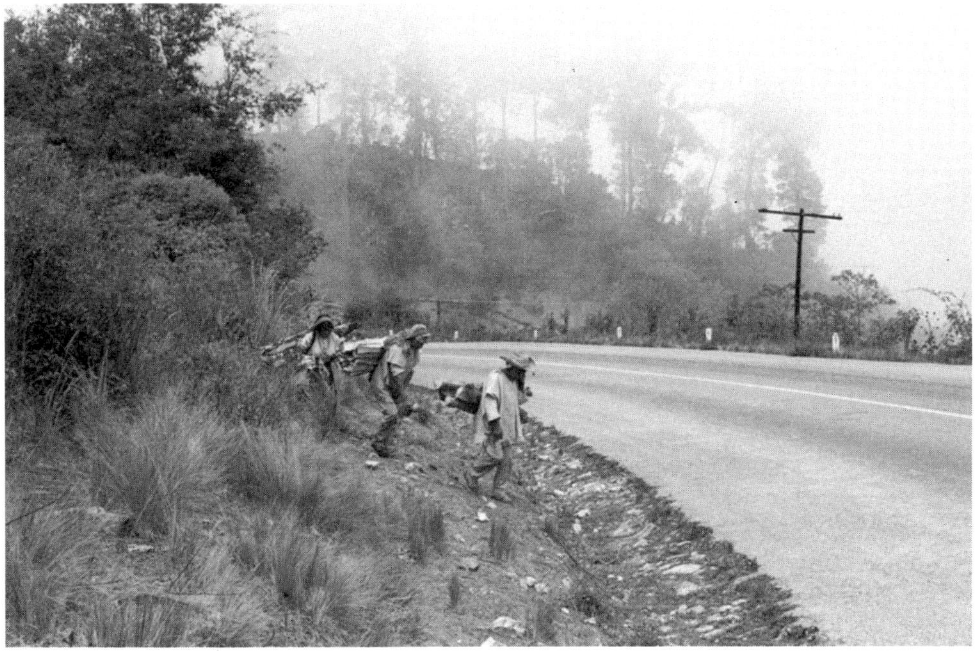

This 1971 photograph of people crossing the Pan-American Highway while they gather firewood in Chiapas, Mexico, hints at the landscape's local meanings and values. (Frank Cancian Papers, MS-F034, Special Collections and Archives, The UC Irvine Libraries)

assumptions about US superiority that underpin their entire driving enterprise. The camaraderie that sometimes emerges between overlanders and mechanics when their vehicles break down due to road conditions or the wear and tear of long distances is but one way that the Pan-American Highway offers potentially transformative connections. These may open traveler's experiences to the deep webs of relations and histories that make up the local landscapes through which they travel. Overlanders are connected to these landscapes, after all, through decades of US interventions in the region.

Overlanders drive the Pan-American Highway like US Americans drive through natural areas. The technology of their vehicles, outdoor gear, and communications systems allows them to enjoy landscapes through a kind of leisure travel informed by US ideas about wilderness deeply embedded in national identity. In this way, overlanders extend a very specific and by no means universal way of seeing and knowing the landscape. This is clearly observed in their photographs and travel writing. But overlanders' effects in the world are not limited to representations. Their vehicles leave tracks in the dirt. They buy things from local people. Their technology enables a kind of freedom, but also leads to missed connections. If the Pan-American Highway was once imagined by North Americans as the main street through a hemispheric neighborhood, for overlanders at the beginning of the twenty-first century driving the Pan-American Highway highlights racial and cultural differences between North and South without any pretense of uplift. Without discourses of progress, overlander travel accounts reveal how the awe-inspiring experience of driving the length of the Pan-American Highway is built on assumptions of US superiority that inform a long history of political and economic interventions in the region.

Inconclusive Returns, Pan-American Possibilities

If Pan-American Highway road trips are about the experience rather than the place, how do overlanders fare as people? According to their blogs, their journeys of personal transformation are as inconclusive as their narratives of return. One person's account stands out as an example of what a successful road trip might look like. Gushing about how it has been the best year ever, they express no regrets about pausing their career and having to start over one day. Whether they are able to start over after returning is an open question. The blog does not reveal what happens next. Other overlander blogs refuse to end, in different ways. Some conclude their pan-American journey poised to begin a new transcontinental road trip on the other side of the world. Others simply stop blogging in the middle of their trip. And some write about a return in which they begin or continue to pursue might-be-projects as self-employed entrepreneurs. Maybe there is no way to write a return to late capitalist life because in fact they never escaped.

Technology on the road offers overlanders temporary respite from alienation by helping them connect with nature and with other overlanders. These connections, however, also distance overlanders from the people and places they encounter. Traveling with less technology, however, may not in and of itself help overlanders look beyond their personal experience to better grapple with the historical violence that makes their road trips possible and how this violence continues to shape their experience of Latin American landscapes. Regardless of how much or how little gear overlanders haul across the North and South American continents, the road is always there. And while the Pan-American Highway idea leads overlanders to the wild places they yearn for, the road itself leads to much more. The road leads to roadblocks, to mechanic's garages, to places in between the sights pursued by overlanders. It leads to opportunities to learn about how travelers from the United States are connected to the places they visit in ways that may not yield a beautiful photograph or good feelings, but that may reveal shared histories.

NOTES

1. This analysis of overlander travel writing is based on a close reading of twenty blogs whose start dates range from 2006 to 2019. Initial data collection was conducted in 2014 and updated in 2015 and 2019. While Canadians, Europeans, Latin Americans, and others also produce travel writing about their experiences overlanding the Pan-American Highway, this study focuses on US Americans because of the deep involvement of the United States in the Pan-American Highway project from its inception. Analysis of US travel blogs focused on the following themes: stated identity, funding for travel, stated aims and reflections on achieving those aims, cultural references, and representations of local cultures and landscapes. The author is grateful to Massimo Moraglio, Melina Piglia, and Dhan Zunino Singh for commenting on earlier versions of this paper.

2. Robert B. Riley, "Vision, Culture and Landscape," in *The Camaro in the Pasture: Speculations on the Cultural Landscape of America* (Charlottesville: University of Virginia Press, 2015), 138–45.

3. Robert B. Riley, "Auto Territoriality," in *The Camaro in the Pasture*, 8–10.

4. Vanessa Watts, "Indigenous Place-Thought and Agency amongst Humans and Non-humans (First Woman and Sky Woman Go on a European World Tour!)," *Decolonization: Indigeneity, Education & Society* 2, no. 1 (2013): 20–34; Enrique Salmón, "Kincentric Ecology: Indigenous Perceptions of the Human-Nature Relationship," *Ecological Applications* 10, no. 5 (2000): 1327–1332.

5. Patrick Wolfe, "Settler Colonialism and the Elimination of the Native," *Journal of Genocide Research* 8, no. 4 (2006): 387–409.

6. Warren James Belasco, *Americans on the Road: From Autocamp to Motel, 1919–1945* (Cambridge: MIT Press, 1979).

7. Gijs Mom, "Orchestrating Automobile Technology: Comfort, Mobility Culture, and the Construction of the 'Family Touring Car,' 1917–1940," *Technology and Culture* 55, no. 2 (2014): 229–325; Paul Sutter, *Driven Wild: How the Fight against Automobiles Launched the Modern Wilderness Movement* (Seattle: University of Washington Press, 2015).

8. On wilderness and the frontier myth, see William Cronon, "The Trouble with Wilderness: Or, Getting Back to the Wrong Nature," *Environmental History* 1, no. 1 (1966): 7–28.

9. Timothy Robin Nicholson, *Five Roads to Danger: The Adventure of Transcontinental Motoring* (London: Cassell, 1960).

10. Gordon Pirie, "Non-Urban Motoring in Colonial Africa in the 1920s and 1930s," *South African Historical Journal* 63 (2011): 38–60; Nicholson, *Five Roads to Danger.*

11. Georgine Clarsen, "Tracing the Outline of Nation: Circling Australia by Car," *Continuum* 13, no. 3 (1999): 359–69, and "Automobiles and Australian Modernisation: The Redex around-Australia Trials of the 1950s," *Australian Historical Studies* 41, no. 3 (2010): 352–68.

12. Kurt Moser, "The Dark Side of 'Automobilism,' 1900–30: Violence, War, and the Motor Car," *Journal of Transport History* 24, no. 2 (2003): 238–58.

13. Georgine Clarsen, "Machines as the Measure of Women: Colonial Irony in a Cape to Cairo Automobile Journey," *Journal of Transport History* 29, no. 1 (2008): 44–63; Sasha Disko, "'The World Is My Domain': Technology, Gender, and Orientalism in German Interwar Motorized Adventure Literature," *Transfers* 1, no. 3 (2011): 44–63; Jakob Krais, "Mastering the Wheel of Chance: Motor Racing in French Algeria and Italian Libya," *Comparative Studies of South Asia, Africa and the Middle East* 39, no. 1 (2019): 143–58; Lewis H. Siegelbaum, "Soviet Car Rallies of the 1920s and 1930s and the Road to Socialism," *Slavic Review* 64, no. 2 (2005): 247–73; Tal Zalmanovich, "'Woman Pioneer of Empire': The Making of a Female Colonial Celebrity," *Postcolonial Studies* 12, no. 2 (2009): 193–210.

14. Georgine Clarsen and Lorenzo Veracini, "Settler Colonial Automobilities: A Distinct Constellation of Automobile Cultures?" *History Compass* 10, no. 12 (2012): 889–900.

15. Thomas McCormick, "From Old Empire to New: The Changing Dynamics and Tactics of American Empire," in *Colonial Crucible: Empire in the Making of the Modern American State* ed. A. McCoy and F. Scarano (Madison: University of Wisconsin Press, 2009), 63–82.

16. Lars Schoultz, *Beneath the United States: A History of U.S. Policy towards Latin America* (Cambridge, MA: Harvard University Press, 1998), 86–90.

17. Benjamin A. Coates, "The Pan-American Lobbyist: William Eleroy Curtis and U.S. Empire, 1884–1899," *Diplomatic History* 38, no. 1 (2014): 43–44; Ricardo Salvatore, "The Enterprise of Knowledge: Representational Machines of Informal Empire," in *Close Encounters of Empire: Writing the Cultural History of U.S.-Latin American Relations* (Durham, NC: Duke University Press, 1998): 76.

18. Salvatore, "Enterprise of Knowledge," 81–86.

19. Rosa E. Ficek, "Imperial Routes, National Networks and Regional Projects in the Pan-American Highway, 1884–1977," *Journal of Transport History* 37, no. 2 (2016): 129–54.

20. María del Carmen Collado, "México y Centroamérica en la formación de la política de la Buena Vecindad," *Latin Americanist* (March 2010): 51–70.

21. Amy Spellacy, "Mapping the Metaphor of the Good Neighbor: Geography, Globalism and Pan-Americanism during the 1940s," *American Studies* 47, no. 2 (2006): 41.

22. Spellacy, "Mapping the Metaphor," 43.

23. Catherine Coblentz, *The Pan American Highway* (Washington, DC: Pan American Union, 1942).

24. Harry Alverson Franck and Herbert Charles Lanks, *The Pan American Highway from the Rio Grande to the Canal Zone* (New York: D. Appleton, 1940); Sullivan C. Richardson, *Adventure South: Three Men and a Lone Car Blaze the Pan American Highway Route down Two Continents to Cape Horn!* (Detroit: Arnold-Powers, 1942); Paul Pleiss, "A Motor Journey Round South America," *Geographical Journal* 101, no. 2 (1943): 66–78.

25. James J. Flink, *The Automobile Age* (Cambridge, MA: MIT Press, 1988), 27–28.

26. Cotton Seiler, *Republic of Drivers: A Cultural History of Automobility in America* (Chicago: University of Chicago Press, 2008).

27. Franck and Lanks, *The Pan American Highway.*

28. Ernst A. Jahn, *Latin American Travel and Pan American Highway Guide* (New York: Compsco Publishing Company, 1966); Pan American Union, *Motoring in Central America and Panama: A Compilation of Information on the Pan American Highway from the Mexico-Guatemala Border to the Border between Panama and Colombia* (Washington, DC: Pan American Union, 1961); Pan American Union, *The Pan American Highway System: A Compilation of Official Data on the Present Status of the Pan American Highway System in the Latin American Republics* (Washington, DC: Pan American Union, 1969).

29. Pan American Union, *The Pan American Highway System.*

30. Erica Victorson, "The Lost Coast of Michoacán—Part 1," *Song of the Road,* January 14, 2014, http://songoftheroad.com/2014/01/lost-coast-michoacan/.

The Peace War. Digital painting by Stephan Martiniere. (Reproduced courtesy of Stephan Martiniere, Green Monkey Design, LLC)

VI. Reflections on Pasts and Futures

2015

The Goose and the Dish

ROBERT B. RILEY

Is it so bad, then, to be a tourist? I would hope that by now that longstanding and condescending dichotomy between travel and tourism has disappeared. I never did understand why bad water, spoiled food, multitudes of insects, thieving merchants, and smelly, dangerous, overcrowded buses were requirements for an "authentic" landscape experience. One is either an insider or one is not. J. B. Jackson, reflecting back upon decades of teaching, concluded that his purpose had been that of producing intelligent tourists. All right, tourists then. But what kind of landscape will we be experiencing? A very different one, I believe.

A clue can be seen in our own not urban American landscape: a richly productive agriculture overlaid by the homes and proliferating appurtenances of those who see the land primarily as a residential amenity. It is produced not only by residents who have no ties to agriculture but by other new population types and attitudes: high-tech recreation that views nature as only an exciting stage set; agricultural workers who commute, just as city dwellers one time did, miles a day from their residence to work eight hours and are paid with checks from a bank from some remote state on the account of Happimeadows AgriSystems, Inc.

I have noted before the initial impact of the automobile, propane tank, and septic tank. Then came satellite television with its attendant minion VCRs, one more amenity to enjoy without locational restrictions. The Internet after the turn of the century has been a major shaper of the new landscape, not only allowing more people but enticing creative people, formerly dependent on conventional urban infrastructure and face-to-face meeting. All this is more than a new demography, a new density, and a new habitation settlement and transportation pattern.

The new landscape relies heavily on privatization of services and seems to obey no laws except those of the market place. Nationally franchised dealers handling propane, fertilizer, herbicide, and equipment rental anchor the new landscape. Placement of the human artifacts on the land, even the use of the land, sometimes seems arbitrary. We have a traditional image of what the landscape should look like, an image based on real agriculture landscapes in which use and habitation followed certain patterns. Even if the depth and causes of the patterns and their interacting complexities were not visible or even known by the outsider, there was still a sense that some form of order was operating, that there were coherence and a logic rather than arbitrariness. We make sense of our exterior world by vision, and if there is no apparent visual order to this landscape, how are we to understand it? How are we to feel other than ill at ease in it? The problem is that this new landscape is in large part aspatial.

Our traditional landscape was understood in terms of a sophisticated (if seldom useful) regional science, central place theory, hierarchical settlement, agricultural land rent, highest and best use, and so on. But the new landscape is a *network*, based on different motivations, economics, and sociology. It is a network with many fewer locational, more spatial distance restrictions than the old network and, in fact, with the electronic communications about as aspatial as any phenomena in space could be. We have not developed any theoretical knowledge from this network yet, and such models that come will be vastly different and much more complex, and less spatial even, than those we remember.

How are we to understand the relationship between this new network and the traditional agriculture pattern? It's easy to think of the new landscape as seeping out of the city along the interstates and inundating the weakest cracks in the old landscape, as if it were some spreading disease. Easy, but not useful. If we were to use an epidemiological model, we would better think in terms of a widespread airborne spore dispersion in a time of global travel and transport, for the new landscape seems to land in unexpected spots and form few predictable clusters, tentacles, or centers. The most useful view of the new landscape is as a *very* loose net laid down over, but almost independent of, the old landscape. Interaction between the two takes place at infrequent and unpredictable junctures, or nodes or linkages.

This is the best model of globalization of the first century of the new millennia: a vast network of new technologies and communications draped loosely over a world landscape still largely agricultural or extractive. Travelers and anthropologists have noted the concurrent existence of these networks in less developed areas of the world. In the tropical rain forest, for example, a network of river travel based on the outboard engine lies over an older trade network of paths and the occasional road. DVD players dot jungle and desert along the network of electrical transmission lines. In our new American landscape its essential services seem less those of bank and grocery store than beauty parlor and video rental outlet. Few of us have even recognized this new landscape yet, much less begun to understand it. Certainly no scholars, planners, theoreticians predicted its coming, although Jackson did forecast some of its characteristics: mobility, change, sensory rewards, and transience.

We now inhabit two differently organized landscapes, one overlaid upon the other. Trailer owners change the holiday costumes on the concrete goose that stands in the front yard next to the satellite dish. The major engine driving change is money. Money sees the land and its development as product. This is true, in different guises, in socialist countries and capitalist countries alike. Nowhere is it more blatant than in the developed world. Lenin asked, "Who profits?" The answer is not easy to find; profits are passed through levels upon levels of financing and manipulation and, unlike the millionaires' mansions of past times, are not discernible by simply examining the visible landscape.

By current indications, the landscape ahead of us is unlikely to be of ecological soundness, or of justice, or of equity. How ironic that contemporary design, at least at its high end, celebrates this world. If it lacks the unity of earlier styles, maybe that is for the better, or at least more appropriate for our time. Indeed, that might be the best we can hope for in the near future: patches of beauty, of brilliant splendor, scattered glowing gems over a global landscape of privilege and power, of disorganization and disenfranchisement. Not a landscape of dark, satanic mills amid a green and pleasant land of smallholders, but the crystal and concrete information nodes of Bangalore or Karnataka rising above urban squalor and high-technology agricultural plantations. Earth may abide forever, but not the earth we know.

The Environment Is a Public Good

ACHVA BENZINBERG STEIN

Whenever we met, Bob and I would get into serious discussions about landscape architecture and who were landscape architects. Bob was thoughtful and articulate, solidly based on his American background, and trained as an architect. Although he was not raised in a typical middle-class neighborhood, the world in which he lived was one where most gardens were attached to houses on private land. Historically, the United States always treated the land as a commodity to be sectioned off and given to individual pioneers just to get it settled and cultivated. Later, it became an article of commerce to be bought and sold like any other form of property or possession. True, some land has been reserved for the public domain—streets, parks, and other federal, state, and municipal uses—but the vast majority is in private hands. The ownership of land is the primary relationship that most people have with their environment.

Growing up in a "socialist/pioneer" society, my background was defined by Israel of the 1940s and 1950s of the last century, where most work was focused on "building the country and restoring the land," a national community effort in which everyone participated. Following the prevailing ethos, almost all the land belonged to the nation, managed through its public institutions, including the government, the municipalities, and the Jewish National Fund. The buildings were on plots leased for either forty-nine or ninety-nine years but not owned by the occupants. Even the farms and *kibbutzim* (utopian agricultural communities based on socialist principles) were on state-owned lands, managed in trust, and supported by a vast public enterprise of scientific and technical institutions in which the entire nation took pride. There were no private gardens. Instead, there were public gardens that were shared by all the surrounding residents.

I came to the United States to study landscape architecture. The direct translation of the term *landscape architecture* in Hebrew is "planning and design of scenery." In my mind landscape architecture was not about garden design but about the land, public areas, and the relationship between these and the broader landscape.

Bob and I spent time at each of our meetings discussing the differences in our understandings of the role of the landscape architect in our respective worlds, challenging each other in the hope of coming up with a common set of dimensions to encompass both of our views.

Our conversations took on a special meaning because our camaraderie meant we could frequently discuss quite opposed viewpoints in a constructive and friendly environment. I deeply miss them.

For Whom Are We Working?

In his short essay "The Goose and the Dish," Bob suggests that perhaps "the best we can hope for in the near future: patches of beauty, of brilliant splendor, scattered glowing gems over a global landscape of privilege and power, of disorganization and disenfranchisement. Not a landscape of dark satanic mills amid a green and pleasant land of smallholders, but the crystal and concrete information nodes of Bangalore or Karnataka rising above urban squalor and high-technology agricultural plantations. Earth may abide forever, but not the earth we know."[1]

I objected to this view of the inevitable future of the landscape of the world. I argued that this interpretation was shaped by the belief that the current political system would continue to be ruled by a market economy. I argued that the reason for the landscape's present appearance is the short-term profit-oriented political system and the overwhelming increase in population. I ascribed the current misuse of land to the commodification of land supported by an increasingly corrupt leadership that almost completely ignores the growing environmental crisis. This is part and parcel of manipulating the public mind and curtailing democratic norms, which has led to a lack of funding to combat ecological calamities and maintain (or maybe slightly moderate) population size.

At the base of all of Bob's and my discussions was our mutual understanding that the proper role of human behavior on this earth is the sustainable and resilient use of what nature has made available. My personal conviction is that the actual client, regardless of who paid for the services, was the public. Our professional work had to have the public good always in mind.

But what is "the public good"? And what are our responsibilities both as professionals and as members of our communities?

In a technical sense, a "good" is viewed, within this context and following some dictionary variants, as having, or generating, two essential qualities: first, it is tangible in the sense that it is capable of being treated as a fact, or understood and realized; and second, it has intrinsic value in terms of relating to the fundamental nature of a thing. It is ethically neutral with respect to its "good" effect on society, although that also is usually presumed to be positive. It also excludes money.[2]

The issue of "public good" was first discussed at length by the father of modern economics, Adam Smith. While it is not the place of this article to engage in a discussion of the finer points of his argument, it should be sufficient to note that Smith felt that society should be responsible for erecting and maintaining those public institutions and those public works that, though they may be in the highest degree advantageous to a great society, are, however, of such a nature, that the profit could never repay the expense to any individual or small number of individuals, and for which it cannot be expected that an individual or a small number of individuals should erect or maintain.[3]

In the nineteenth century, the idea was put forth that the objective of any economic system and role of government was to improve the well-being of the individual, not as an isolated person, but as a member of community and society. These ideas were first developed by a group including Adolf Wagner, and were then brought to the United States by two well-known immigrant professors, Richard Musgrave and Gerhard Colm.[4] They were then generally accepted in the United States until the 1950s when the economic theorist Paul Samuelson wrote his seminal article on the subject.[5] Samuelson defined a public good as one that exhibits one or both of the following characteristics: non-excludability, meaning goods that are difficult or impossible to keep nonpayers from consuming, and nonrivalry, meaning that the consumption of the good by one individual does not reduce the availability of the good for consumption by others. However, this definition makes the public good or public goods economic negatives, representing "market failure," goods that the market cannot or will not produce.

John K. Galbraith, writing in *The Affluent Society* (1958), also believed that public goods (many products and services—libraries, schools, roadways, drinking water, etc.) are "things [that] do not lend themselves to [private] production, purchase, and sale. They must be provided for everyone if they are to be provided."[6]

By the latter half of the twentieth century, most economists had negated the concept of a "public economy" and replaced it with the idea that societies operate via markets. The idea of "public goods" came to be viewed as a "market failure," giving it an essentially negative connotation. This remained until very recently when opposing views began to call for a reassessment.

Two British economists, David Hall and Tue Anh Nguyen, recently calculated the economic benefit of public services. They reported that public sector activity, directly and indirectly, supports half the formal jobs in the world, and has a comparative advantage over private contractors in delivering public goods such as universal access to healthcare, affordable housing, and protecting the planet from climate change.[7]

Today we differentiate between "public goods" and "social goods": Public goods are produced by public sector agents of the polity—government agencies, public authorities, public universities, etc.—not by businesses, civil society, NGOs, households, or individuals. Goods produced by these latter entities may be enjoyed by the public and may be called "social goods," but they are not public goods. The distinguishing characteristics of public goods are that they are created through collective policy choices (voting) and are paid for collectively (public financing).

The importance of managing, designing, and monitoring the physical environment as a public good is based on the rationale that public services are, by and large, discretionary. Their creation and continued existence depend on the choice of the agents or organizations. It also depends on the volunteers' devotion and commitments, subject to the control of those who provide the service and the financial backing. Therefore, public services are neither excludable nor rivalrous. On the other hand, government agencies are controlled by the public, and their services are subject to citizen scrutiny. They can impose policies "to limit pollution, restrict resource exploitation, or create the right incentives to promote or protect environmental quality . . . and promote economic efficiency . . . for which the societal benefits exceed the costs." On international and global levels, public goods vs. public service issues become more acute since coordination and monitoring require sovereign nations to comply with the international agreements. There is no stable and recognizable body to impose restrictions (except the United Nations and its various associations). The world public has no power (save on the national level) to accept or oppose the policies.

Historical Changes Affecting Professionals

Learning skills and acquiring knowledge have also gone through several mutations in the last several centuries. The source of transmitted knowledge was transferred from home and family to individual master craftsmen and later to the guilds. Each stage came with a moral code that changed and evolved based on the services provided and the society's culture. After the Industrial Revolution, when large masses of capable workers were needed, teaching organizations, such as schools, universities, or academia, took over. They professed to base their knowledge on science, empirical investigation, altruism, and moral worth. This was the educational background of the pioneers of the landscape architecture profession.

They assumed that the government's goal (at the time) was dedicated to improving the well-being of both the individual and the society at large. They believed that their choice of profession would benefit the public.

Frederick Law Olmsted (1822–1903) introduced the term "landscape architect" and was the first to be known to the public by this new name. Luckily, he was actively involved in working for the public good and set up an impressive legacy for the future generation to follow. His work encompassed a wide range of projects for the public good—urban parks and master plans for housing developments, and the design of transportation projects, among them the Eastern Parkway in Brooklyn. However, planning and designing the American landscape really started with other pioneer landscape designers. Among these were Harland Bartholomew (1889–1969), who prepared city plans for Newark, NJ, (1911) and St. Louis (1914) as well as a significant portion of the report of the Interregional Highways Committee laying the groundwork for the American interstate highway system; and Earle Sumner Draper (1893–1994), who prepared more than three hundred subdivisions, college campuses, cemeteries, as well as plans for one hundred mill towns and parks and was the first planner for the TVA and the Morristown greenbelt. Daniel Ray Hull (1890–1964), together with Frederick Law Olmsted Jr. (1870–1957), made a comprehensive survey of ecologically valuable lands in California that was directly responsible for establishing California's extensive state parks system.[8]

Landscape architects have designed and supervised comprehensive plans for cities, counties, regions, state parks and systems, subdivisions, housing projects, highways, and even sanitation and street cleaning regulations. Since most of this work was either controlled or financed by public agencies, it must also be noted that the quantity, content, and style vary with changes in the political atmosphere.

The understanding of the havoc which human beings have inflicted on the world has been recognized since antiquity. The increased scientific awareness in the nineteenth century could be found in the coining of the term "ecology" by Ernst Heinrich Philipp August Haeckel in 1866[9] and Henry David Thoreau's declaration that "in Wildness is the preservation of the World" in 1863. The term "acid rain" was introduced by Robert Angus Smith in 1864 in his book *Air and Rain*.[10] In 1864, George Perkins Marsh published the first systematic analysis of humanity's destructive impact on the natural environment.[11] The first national park in the United States, Yellowstone National Park, was established in 1872 and was followed by the formation of Yosemite and Sequoia National Parks.

The early twentieth century brought into public focus issues of environmental destruction, mainly as the understanding grew of the failure of farmers to see the conflict between quick profit and the laws of nature, which became apparent with the ecological and human disaster of the Dust Bowl.

Between the 1920s and the 1940s, progressive architects in Europe and the United States were convinced that Modernism would resolve the conflict between the environment and the social order with appropriate technology. Many experimented with materials better suited to "mass housing," and they worked in various capacities on public housing and suitable workers' homes for large projects such as the TVA dams and the Farm Administration's programs for housing migrant agricultural workers. Serge Chermayeff articulated the mood of many design professionals when he said:

> Nothing needs occupy our attention seriously at this particular time in examining new materials other than those which have direct social significance. The architect's fullest attention and best service are the practical and economic problems of housing, industry, and trans-

port. The erection of new buildings in a manner and with material which will not saddle the absoluteness of the bygone on ourselves or those who will come after us.[12]

In the United States, with substantial subsidies, the Works Progress Administration (WPA, 1935–43) employed thousands of architects, engineers, and landscape architects who chose to work for the public good rather than devote themselves to building private homes and gardens for the well-to-do. When the federal government stopped the WPA and redirected the money to other relief agencies such as the Civilian Conservation Corps (CCC), which employed more people, the profession's conviction of working for the public was continued with the assistance of states and cities. However, a changed government financing model and the large-scale investment of private funds after World War II created a vast enterprise of developing segregated suburbs, made up of residential single-family structures and an extended network of roads for private transportation. Returning soldiers and families migrated to the suburbs in great numbers, and the automobile displaced the urban centers and the urban public realm.

Historically, many landscape architects and designers have considered their occupation to be primarily that of garden designers for the well-to-do. The short period of commitment to projects for the public good in the first part of the twentieth century ended with the period of large-scale suburban creation. From the 1950s and up through the 1980s, many landscape architects returned to concentrating on designing gardens, both small and large. However, they also designed resorts, shopping centers, amusement parks, and golf courses. They helped developers to erect townhouse types of world dwellings in "Planned Unit Developments" that were not intended to house struggling populations, and beautification projects became bread-and-butter jobs for many offices. Riley called the results "a part old, part new physical vocabulary." He continued, describing the vocabulary as "the eternal bedding plants now sitting in designer boxes, some tastefully minimal sign regulation, streetlights reminiscent of the 19th century, designer approved bicycle racks, a variety of paving materials (hardscape) suggesting brick and cobblestones and cast-iron grates for the trees."[13]

The image of the blue earth as seen by the astronauts on the moon in 1969 entered the nation's psyche. The fragility of our "spaceship" made many all over the world notice that this was the only known place in the galaxy in which humans could survive and that if we wanted to continue living, we needed to drop our complacency and learn to conserve our resources.

Although the Flood Prevention (1954), Clean Air (1963), National Wild and Scenic Rivers (1968), and the National Trail System (1968) acts had been enacted by the government, the concerned public—especially the young—did not see any major movement to clean the existing pollution and, in general, the careless treatment of the earth and all the living and the nonliving things on it continued as before.

Earth Day was first celebrated in April of 1970 under a call for "the fight for survival" and launched the environmental movement. Additional waves of environmental legislation were passed. The progressive part of the profession looked toward working for the public good with much hope. Ian McHarg's book *Design with Nature* was published in 1969 and was influential in refocusing the profession's attention on the difference between the way we had been taught to plan and design the landscape and the way we needed to revise our working processes.[14] The book has gone through more than twenty-five editions and has been translated into many languages. It has also reached a large audience of laypersons in addition to the architecture profession.

During my student years in the 1960s, a renewed commitment took over the younger generation and the progressive elements of the profession in Berkeley. The state of the environment and identifying public needs dominated discussions among students. Rachel Carson's seminal book, *Silent Spring* (1962), was already part of our required reading and the graduate students read Stewart Udall's *The Quiet Crisis* (1963), his manifesto on environmental pollution. Fellow students were active in promoting the formation of Redwood National Park in northern California. Some were also concerned about the lack of solid programs to conserve the Sacramento Delta.

Many of us absorbed these concerns, and out of this ferment came the beginnings of a new activism that was directed toward our role as creators of the human environment, as we saw it. "People's Park" and "People's Architecture," among many other protests, had grown out of a merger of the movements to protect both the earth and the urban environment, coupled with an awareness of the massive environmental destruction in Vietnam. But the university's College of Environmental Design faculty were only marginally involved in this unrest in Berkeley. One exception was when Garrett Eckbo, then the chair, turned the Department of Landscape Architecture facilities over to the students to create protest materials around the invasion of Cambodia in 1970, causing him to be summoned to Sacramento by then-governor Ronald Reagan.

When we graduated, many of us started out by looking for public employment. We felt strongly that we should contribute to the public good, using our acquired skills. The Bureau of Land Reclamation, the Fish and Wildlife Agency, the transportation departments, both state and federal, and regional, governments opened their doors to us. Many landscape architects started their careers in these public agencies, working on much-needed public landscape issues.[15]

We were less enamored of the idea of private gardens and did few of them during our professional working days. The Vietnam War continued to shadow our life for a while, and the political ideology of the federal government worked consciously to rule and divide, influencing the public around us. Our activism subsided to mostly individual and small group contributions and several academic programs with specialized community assistance courses. In the 1980s, to control environmental proposals before they became public, President Reagan cut the budget of the Environmental Protection Agency twice, first in 1981 and again in 1984, with the cuts amounting to 44 percent of the 1980 level. The EPA's staff was reduced by 29 percent, and its budget was reduced to 44 percent of its 1978 level. The number of enforcement cases submitted to the EPA during that period declined by 56 percent. The solar water heating system on the White House roof, installed by President Carter, was dismantled as a symbolic act.

Altruism and the ethic of working for the public good had been the ideals for many professionals up to the middle of the twentieth century in the United States. Most professional design work was carried out in private offices. These offices were headed by almost mythical professionals. Among them were Lawrence Halprin, Dan Kiley, Peter Walker, and Hideo Sasaki, supposedly autonomous wealth generators and supported by private clients connected to the political elite, whose essential requirement was esthetic design expressions on the land. These prerequisites frequently conflict with the ethics of the ecological management issues our world has been facing. By the end of the century, landscape design practice had largely shifted to corporate organizations. It was carried on under the excuse that corporations were better equipped to carry out a full range of work: from site design to regional planning to construction documents. But the tax structure and the laws which treated corporations as separate entities exempt from personal litigation were the real motivations. In addition, as noted by Laurie Olin, "these firms had the ability to interact with great varieties of clients including corporate investors, developers, and public agencies because they are the social peers of these clients."[16] These professional design corporations were supported by the ultra-wealthy establishment and their conglomerate's companies. Additionally, countries like Saudi Arabia, Iran (under the Shah), and, for the last thirty years, China, and other authoritarian governments around the world opened up to large international design firms. These design corporations were hired to design mostly luxury projects, where the goals appeared to be personal glory and the enhancement of the status quo. Projects for the public good that attempted to improve the quality of life for the lower economic strata of society or to contribute to efforts to deal seriously with the climate changes affecting the planet were sponsored mainly by the World Bank and similar international organizations. These projects were carried out by engineering and planning firms. On occasion, these firms hired a landscape architect to work with them, and this collaboration with engineers and planners often produced favorable results.

We entered the twenty-first century recognizing the fact that the political powers controlling most countries, regardless of their position on the political spectrum, whether left, right, or center, are at best barely curbing this destruction, and are often even actively supporting and encouraging the continuation of the current exploitation and exhaustion of available resources. At the same time, many worldwide realize that they need to start taking matters into their own hands. Nongovernmental organizations (NGOs) such as the Sierra Club, altruistic organizations like 350.org, and a number of wealthy individuals like Bill Gates and Jeff Bezos have taken over the role previously led by governments and have begun working for the social good. They have bought land and passed it to public trusts. They have taken control over wildlife protection. They have started to use the media to enlighten the public on environmental protection issues and the need for open space. The United Nations adopted the Sustainable Development Goals (2015), which have taken the idea of landscape conservation and turned it into land management. The three-pronged sustainability goals—economy, social issues, and environment—became the focus of "the fight for survival."[17] Many segments of society have voiced their opinions in voting, in writing, and in demonstrations. Examples can be seen not only in developed countries but also in the struggles of Indigenous Amazonian victims of oil spills and pollution, Native Americans fighting against the proposed Dakota Access pipeline and North Dakota's fracking fields, and First Nations of western Canada struggling against the exploitation of the tar sands turning the land into moonscape.[18]

However, as we have seen from the results of the 2016 election, the members of the financial elite have largely held on to their wasteful habits and the deeply entrenched belief that profit is the primary, if not sole, reason for any human activity.

Since the Declaration of Concern in 1966, when ASLA (the American Society of Landscape Architects) president Campbell Miller, *Landscape Architecture* magazine editor Grady Clay, and Ian McHarg wrote, "A sense of crisis has brought us together. What is merely offensive or disturbing today threatens life itself tomorrow. We are concerned over misuse of the environment and development which has lost all contact with the basic processes of nature."[19] Landscape architects have professed that they are the stewards of the land. But not until 2008 did the ASLA publish the statement to show the organization's commitment to work toward climate change mitigation. In England, the Landscape Institute also published a position paper clearly delineating their commitment to the environment with illustrations of sustainable projects. Only in 2017–19 did the ASLA publicly and officially commit to working to secure a sustainable and resilient future, including environmental justice, racial and social equity, and meaningful community engagement.

Concluding Thoughts

What will the coming generations make of the landscape? How will planning and management interact with the need to design? It is not clear.

I believe that the need to manipulate our physical environment and leave our personal mark on the landscape is only partly inherent in human nature. The culture of humanity is still controlled by primal fears and the presumed necessity to dominate the elements. This fear and the actions that result from it have been restrained to a limited extent by the latest trends in education and training. Unfortunately, our religious and ethical belief systems are lagging.

Nonetheless, one issue is clear to me. The scale of design on the land needs to be clearly articulated and curbed. The land is not a canvas to draw on, and the ecological systems are delicate. Many have already been disturbed with consequences that we have recently just begun to understand better, especially the forces of bacteria, fungi, and other pathological agents.

The function of gardens, parks, even urban streets, is mainly geared toward "Ensuring of healthy lives, promote wellbeing for all at all ages, and making cities and human settlements inclusive, safe, resilient and sustainable."[20] These projects do require landscape professionals to deal with the environmental and socioeconomic issues of the areas the projects are located in. But the change in scale and complexity of issues involved in landscape planning and design transform, alter, and distinguish it from garden design.

Landscape planning and design deal with complex systems that include, among other issues, the process of decision-making between the public, private interests, business, governmental agencies, and professionals. They also need to advocate and consider all the living and nonliving entities required to maintain life on our planet.

For this, a completely different approach to our work needs to be set in motion. We will need to work in multidisciplinary modes, not only with the allied professions but also with those who can augment our understanding outside of our field. At the same time, we need to expand the range of our activities and contributions.

The question of the scale of our work is often moot—and we need to learn to examine design problems on many levels. The goals and the aims of our work will determine the nature of the team, while the interactions with the environment will determine the appropriate scale.

The growing use of technological innovations and the resulting frame of mind which had taken over our educational system in the twentieth century, an emphasis on function and physical needs, seems to take the front seat in every design discussion. This is not a necessity. Discussions and solutions to design problems need not exclude, nor negate, the need to fulfill our search for order, proportion, rhythm, and other principles in the category of "Beauty and Harmony." They mean only that the creation of the object is not the "final solution" for any project. Instead, the resolution will need to be a changeable entity based on a flexible set of criteria, open to criticism, and multiple revisions over time. In addition to physical alterations of the landscape, we also must understand that much of our activity will necessarily concentrate on research, analysis, and communication.

More than that, landscape planning and design today need to satisfy the seventeen goals of the UN agenda for sustainable development.[21] Eight of these are clearly related to our training, professional knowledge, and skills. Managing and designing the complex systems that we have resolved to do under the professional declaration is not an assignment for either individuals or corporate design firms.

Population growth has significant implications for landscape architects involved with garden design. California alone has six accredited departments of landscape architecture, each training students following the doctrine of "artist and heroic designer," leading their graduates to come out aspiring to leave their personal marks on the land. When Garrett Eckbo (1936) and Robert Royston (1940) graduated from the University of California at Berkeley, there were fewer than five graduate students in their landscape architecture class and about 6 million residents in the entire state. Today there are nearly 40 million residents (with 60 million people expected by 2050), all living and working on the same piece of land. This fact alone should have a decisive impact on the way we train and educate both the designers of the gardens and the landscape planners.

Professionals who work on and with the landscape are generalists. In this, they are unlike scientists, whose knowledge is highly focused on narrowly limited elements and events. However, as landscape architects and environmental planners, we need to have a deeper understanding of the science and logic behind our decisions. We cannot continue to be trained as if we were merely the creators of objects in the way architects are taught. While we are environmental planners and designers, we are not urban and regional planners, and our knowledge base for developing social and political solutions is simply not adequate. Instead, we need to work in teams composed of a variety of disciplines. Learning to work as team members requires education, training, and practice. Unfortunately, we rarely learn or teach this critical skill.

We need to think about teams as "discrete units of performance and not just as positive sets of values." Teamwork is not the same as a working group. A team has to come up with a specific set of purposes and goals. A team combines people with technical or functional expertise, problem-solving and decision-making, and interpersonal skills, and members take responsibility both for the final work and for individual decisions taken by team members on their way to solve a particular problem.[22]

At the end of my career, as I look back at my heated discussions with Bob, I realize that we were mostly thinking alike. The conflict between the near-total commitment to profit at the cost of the environment and the need to treat the land as a public trust held by society became sharper to him. But my world has changed more than his. The "pluralistic" landscape, the "packaged" landscape, the landscape of "nostalgia and hedonism" have spread in various degrees to the rest of the world. They have come about through the dominance of political and economic ideologies and globalization, reinforced by the media.

It is still my belief that the land must not be treated as a commodity. It was given to us in trust to maintain, not to devour and consume. I believe that it is the duty of our profession to help manage the landscape with the help of public institutions, governments, and municipalities.

I hold dearly to the idea that the goods of the environment are tangible. It is a fact to be understood and realized, and it has an intrinsic value. The environment is a "Public Good," and it is not an "economic failure." Rather, it is the responsibility of government and public institutions to conserve, repair, and maintain. Lastly, I believe wholeheartedly that those who teach must base their knowledge on science, empirical investigation, altruism, and moral worth.

I originally came to study *tichnun nof* ("scenery planning"), and I soon learned that the word "nof" was an entirely misplaced term for what we were learning and doing. Put simply, our role as landscape planners and designers does not include the creation of scenery. The landscape we observe is the result of the way we live on the land. We really should be studying and practicing the various ways we can share our space with all the living beings in the universe and manage resources with the goal of resiliency. These goals necessitate using the best ways to resolve the conflict between nature and culture, even provisionally, and learn to adapt these resolutions over time as needed.

As Bob had reframed Erich Isaac's view, I also realized that "the primary relation of people is to people and people's relation to the landscape is secondary."[23] This statement is equally valid under both capitalist and socialist regimes. If we want to affect changes, we need to start at the very beginning of our lives, not only to change our educational systems and political beliefs but also to pass scientifically based, enforceable laws and regulations regarding the conservation and maintenance of the environment. I can only hope that this will occur soon since we are in a fight for the survival of our planet.

We need to recognize and state clearly that we are environmental planners and designers, not just "architects" of the landscape. We need to keep in mind the entire ASLA 2019 Call to Action and actively follow it individually and collectively.[24] Our real clients will forever be the people together with all the living things on earth: our contributions must meet the test of benefiting the planet.

NOTES

1. Robert B. Riley, "The Goose and the Dish," in *The Camaro in the Pasture: Speculations on the Cultural Landscape of America* (Charlottesville: University of Virginia Press, 2015), 156.

2. Dana Dalrymple, "The Role of Scientific and Technical Data and Information in the Public Domain," in *Proceedings of the Symposium Scientific Knowledge as a Global Public Good: Contributions to Innovation and the Economy* (Washington, DC: National Academies Press, 2003).

3. Adam Smith, *The Wealth of Nations* (Edinburgh: Thomas Nelson & Co., 1827), 302.

4. Richard A. Musgrave, *The Theory of Public Finance: A Study in Public Economy* (New York: McGraw-Hill Book Co., 1959).

5. Paul A. Samuelson, "The Pure Theory of Public Expenditure," *Review of Economics and Statistics* 36, no. 4 (November 1954): 387–89.

6. John Kenneth Galbraith, *The Affluent Society*, 40th ed. (Boston: Houghton Mifflin, 1998), 110.

7. David Hall and Tue Anh Nguyen, "Economic Benefits of Public Services," *Real-World Economics Review*, no. 84 (2018): http://www.paecon.net/PAEReview/issue84/HallN guyen84.pdf.

8. Charles A. Birnbaum and Julie K. Fix, eds., *Pioneers of American Landscape Design: An Annotated Bibliography* (Washington, DC: US Department of the Interior, National Park Service, 1995), 16–18, 44–47, 79–83.

9. Ernst Heinrich Philipp August Haeckel, *Generelle Morphologie der Organismen* (Berlin: Verlag von Georg Reimer, 1866), Google Books, https://books.google.com/books?id=dthOAAAAMAAJ&printsec=frontcover#v=onepage&q&f=false.

10. Robert Angus Smith, *Air and Rain: The Beginnings of a Chemical Climatology* (London: Longmans, Green & Co., 1872).

11. George Perkins Marsh, *Man, and Nature, of Physical Geography as Modified by Human Nature* (1864; rpr. Seattle: University of Washington Press, 2003).

12. Serge Chermayeff, lecture to the Royal Institute of British Architects, December 18, 1933, published in *Design and the Public Good*, ed. Richard Plunz (Cambridge, MA, and London: MIT Press, 1982).

13. Robert B. Riley, "The Urban Cosmeticians: Or, The City Beautiful Rides Again," in *The Camaro in the Pasture*, 21.

14. Ian L. McHarg, *Design with Nature* (Garden City, NY: Natural History Press, 1968).

15. I and a fellow classmate were hired by the City of San Francisco to design miniparks under funding from the Johnson administration's America the Beautiful program and I later joined the Association of Bay Area Governments (ABAG) and participated intensely in the preparation of the Open Space Plan for the San Francisco Bay Area—1970–1990, which was adopted and whose results are very much visible today. Some went to work for the Corps of Engineers, and others joined the Peace Corps, while two classmates spent most of their careers with the East Bay Regional Parks.

16. Laurie Olin, "Sociology of Professions: The Evolution of Landscape Architecture in the United States," cited in *Landscape Review* 12, no. 1 (2005): 3–25; email correspondence, 2005 as quoted by Timothy Baird and Bon Szczygiel.

17. United Nations, *The Sustainable Development Goals Report—2019* (New York: United Nations, 2019), https://unstats.un.org/sdgs/report/2019/The-Sustainable-Development-Goals-Report-2019.pdf.

18. Younger and older students from Fiji, Samoa, Vanuatu, Kiribati, Tuvalu, Marshall Islands, Tonga, New Caledonia, the Solomon Islands, and Papua-New Guinea have been demonstrating against the lack of effort to prevent rising sea levels that threaten the very existence of their nations. The poor who endure suffocating air pollution and plastic waste in India, Pakistan, and Nepal, and the native populations and landless workers have been fighting against the salt intrusions and siltation of their wells in the deserts of Africa and the Middle East.

19. Grady Clay et al., "A Declaration of Concern," *Landscape Architecture Magazine,* June 1966.

20. UN, *Sustainable Development Goals Report.*

21. See https://www.un.org/development/desa/disabilities/envision2030.html.

22. Jon R. Katzenbach and Douglas K. Smith, "The Discipline of Teams," *Harvard Business Review,* March–April 1993.

23. Riley, *The Camaro in the Pasture,* 157.

24. American Society of Landscape Architects (ASLA), *The New Landscape Declaration: A Call to Action for the Twenty-First Century* (Washington, DC: Landscape Architecture Foundation, 2019), https://www.lafoundation.org/take-action/new-landscape-declaration#nlddocument.

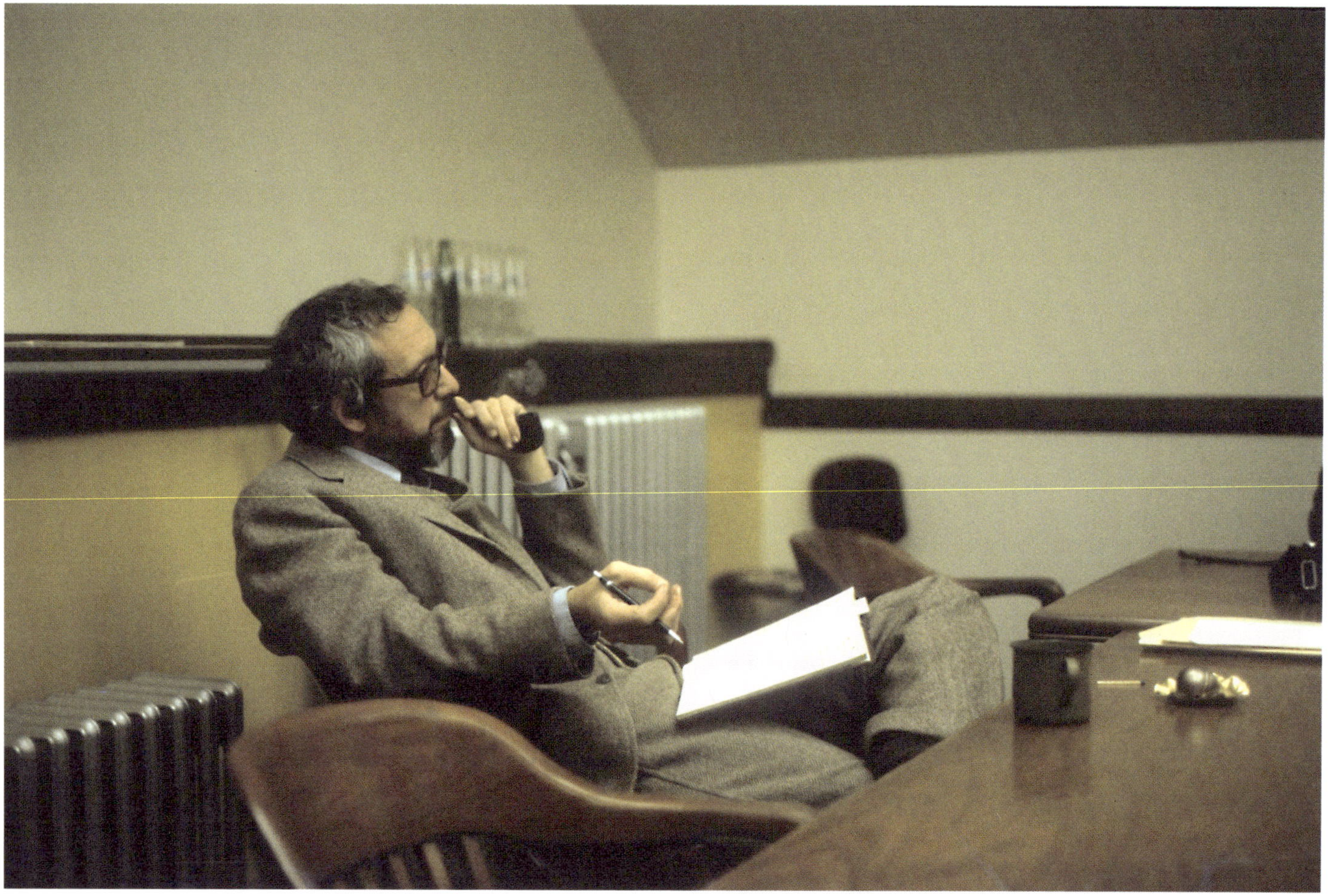

Robert B. Riley at a department meeting, University of Illinois, 1977. (Photograph by Paul Lettieri)

VII. Reminiscences, Anecdotes, Assessments

Michael Van Valkenburgh: Remembering Riley

When my soon-to-be wife, Caroline, and I flew to Urbana to visit the University of Illinois in 1975 just before I enrolled, Bob Riley met us at the airport. In grad school, we all called him Riley. And what a sight Riley was, with a wealth of turquoise jewelry circling his neck and loading his wrists. And, of course, the white Levi's. His vaguely shy face, that day and always, was upstaged by his brilliant eyes and his abrupt and snarky retorts. Riley's ridiculous car made the scene even more striking. It was a long, old-school Cadillac, either inherited from his parents or taken out on loan: I have never been sure. He bragged with irony about its terrible gas mileage. I could not help thinking of the car as an ocean liner aground on the prairie. He did not think for a minute that a gas guzzler was a good idea, just as we know that getting drunk as hell is not a good idea but do it from time to time.

I am pretty sure we had not ended the trip to Illini Union before he told me that what was wrong with our field was too many saps. I took this as fair warning that I might be wise to keep some of my own sentimental thoughts tucked away. Although setting sentiment aside was a big part of the person-environment program that I was entering, his words were a good lesson, nonetheless. Riley was not dissing sentimentality. I am pretty sure he had a large reservoir of it. By any measure, that first hour or so was an awkward meeting, as he was not a fan of the empty words spoken so often during first encounters. And I did not know a landscape that was as flat as Illinois, but I sure knew that I liked him.

Not long after pulling away from the airport, we passed a thick row of Osage orange trees. One could not have graduated from an agriculture school in the 1970s—as I had done—without knowing this tree. But Riley surprised me with a story of their "hog tight" hedgerow planting during World War II, when metal fencing materials were rationed and scarce. Thus, my cultural landscape education began.

Riley and I were both afflicted with "extremely early-to-rise syndrome." We would often run into each other at a diner in downtown Urbana at 6:00 a.m. We would sit at the counter, eating fried eggs, biscuits, and apple butter and talking about everything and nothing. It was there that I learned the most in grad school. Riley hired me to teach drafting and graphics and ironically, its schedule did not allow me to take his graduate course, so I never sat in a classroom with him. The diner was my classroom and the place where we really bonded.

I was assigned to work with Sue Weidemann, one of Riley's friends. Both were brilliant teachers because they asked questions rather than giving answers. I carry some of Riley's lessons with me. He used to say, "You will only be successful if you know what you don't know," and I often share this bit of wisdom with colleagues.

My first semester—ecology and research methods courses—was brutal. To counterbalance some of that, Bob persuaded the faculty to let me fulfill the required design studio requirement by taking classes with the photographer Robbert Flick. While Flick's life's work was just art on the wall to him, it gave me a new way to think about ordering a landscape and its resulting coherence. In other words, coming at landscape-making from its experiential, three-dimensional qualities. Indirectly, it fit with Riley and Weidemann's basic request that, as designers, we imagine how our designs are to be understood by others. I owe all that to Riley.

A couple of years ago I returned to Urbana and spent an afternoon with Riley. We sat in a comfy, informal lounge on campus, and he and his students peppered me with questions. He asked if it would be okay to tell the group that I came from a family of modest means, because so many of his students were from similar backgrounds and would benefit from the reassurance. An older Riley was an even wiser Riley, but I found a happier man, too. Great man, great mind. He stretched every student that he ever worked with. We all know that Bob Riley was a wonderful teacher and thinker, and a wickedly good writer. And yet, the great thing about Bob was he also loved the visual world. He designed his own simple, modern house, and it was a pleasure to be near and in.

He wanted us all to leave grad school more curious about the world and more rigorous and demanding in how we took that curiosity somewhere. It was a privilege to know him in such a formative time, and I loved him a whole lot.

Terry Harkness

Bob wanted to understand landscape and design and so he would seek out and really listen to people involved in architecture, design, and planning. And then he would think about it critically . . . testing, weighing, questioning.

Bob was so unusual in his curiosity and his large view of the world. What I found so refreshing was that he said there are all these related things that are important to think about, or to be aware of—whether you're a landscape architect, whether you're an architect, whether you're a planner—we have to look at the world in a much richer way. This inundated his thoughtfulness as the head of the landscape architecture program and the other things he involved himself in, whether it be *Landscape Journal* or EDRA or what he wrote. He had a much richer and interesting world that was looking outward, not just about what landscape architects thought about or what architects thought about. I just thought that was a kind of magical quality.

Bob was engaged with the people who were thinking about what they were doing and it didn't matter what point of view they were engaged in, but that they were engaged. This was such a startling difference in our program.

He always had a larger view; he was always hungry to know more. I don't know where he got that.

Sharon Harkness

In 1981, after eight years at HOK in St. Louis, Terry was rehired by the University of Illinois. Bob called me to ask nicely if I needed help finding a job. I didn't, but asked him if Terry's job came with a yearly bonus (the Big B), a company car, and spousal travel (knowing of course it didn't). Without missing a beat, Bob said "No, but it comes with a nice metal desk and a phone . . ."

James Wescoat: A Tribute to Bob Riley

Bob Riley was ahead of his time, and of his place. I initially got to know Bob through the Dumbarton Oaks Board of Senior Fellows, where he demonstrated his singular combination of earthy intelligence and frank, down-to-earth decency.

Bob had an open mind to innovative fellowship and symposium proposals. One of the best examples was his support for the Theme Park landscape symposium with Terry Young. The idea was initially greeted with some skepticism; some asked whether this brand of vernacular landscape could be worthy of the world's most venerable research center for garden history? Bob and colleagues prevailed, and by the end of the conference, many participants wondered what landscapes, if any, have *not* been purposely "themed" in one way or another. Bob gave the same creative encouragement and latitude to his colleagues and students.

As I got to know Bob better through Dumbarton Oaks and the University of Illinois, I became intrigued by his attachment to the places where he had lived and studied, beginning with "heady" undergraduate life at the University of Chicago and architecture at MIT (about which I now wish I had asked him many more questions). Bob never lost his love of New Mexico even after decades of commitment to Urbana-Champaign, which became an intellectual center for the field of landscape architecture under his leadership—from landscape history to environmental perception and behavior. Like his close friend and colleague at Illinois, Terry Harkness, Bob loved the land and the department. And he spoke frankly about both, once telling me that "what you see is what you get," which in Bob's view of the world was a strong compliment. I benefited from his experience and advice. Bob mentioned that his appointment as the landscape architecture department head had been greeted with some skepticism, but he persevered and won allies across the university as well as the wider fields of landscape theory and environmental design research.

James Curtis: Remembrance

I had the pleasure of meeting Bob in 1974 during my first week of graduate school and thereafter, as his student and his research assistant—grain elevators! His interests were wide-ranging and he shared them with everyone. He was an amazing person and an inspirational teacher who opened my eyes to the beauty of the vernacular landscape and much else through his introductory landscape course and many subsequent conversations. His teaching and influence remain indelible.

After graduate school I saw Bob periodically in Urbana and elsewhere, very memorably at an open house/tour of the Lovell Beach House on the Balboa Peninsula in Newport Beach. Two memories from school days follow.

> During one session of his introductory landscape course Bob showed and discussed the work of Archigram. During the Q & A period I suggested that the work was fascist. After

a moment of consideration Bob asked if it might serve a different form of totalitarianism. Point taken (and held).

On a winter afternoon in my second year we were in his Mumford Hall office discussing the progress of the grain elevator research as the smell of marijuana drifted down the hall from the Sophomore Design Studio. Bob asked if this smoke was "beyond the limit." He left, put out the fire, and without missing a beat, returned to the subject at hand.

The world is much poorer and less interesting without him.

John Jakle: Bob Riley

I owe Bob Riley a debt of gratitude for helping to further my career as a cultural and historical geographer. When he became chair of the Department of Landscape Architecture at the University of Illinois at Urbana-Champaign (and I a new member of the university's Geography Department), he began to bring to campus many of the nation's leading lights, those not only renowned in park and garden design, but those heavily invested in scholarship given to landscape history broadly defined in terms of built environment. In so doing he followed fully in the scholarly tradition of J. B. Jackson. Earlier, while on the faculty at the University of New Mexico, Bob had helped Jackson edit *Landscape*, the magazine that began to turn the design professions, as well as academic geography, toward appreciation of the architectural vernacular and, especially, common landscapes and places. Jackson, himself, teaching part-time at the University of California's Berkeley campus, had befriended Carl Sauer, the then dean of cultural historical geography. "You are going to Illinois," Sauer had said on the eve of Jackson's first of many visits to the Urbana campus. "Look up that John Jakle. What is he all about?" I had just published several articles that Sauer had read. And thus out of the blue, Bob, whom I had not yet met, arranged a full afternoon meeting with Jackson.

An "Int-er-es-ting" aside (that is how Bob always pronounced the word interesting): I had known of the Riley family for some time. Bob's father and my father had a business relationship. His father spent a career managing large hotels (at least one in Chicago and two in Detroit). And his hobby was model railroading. My father sold insurance and he created and oversaw a group insurance program at Bob's father's second Motor City hotel. As a child I was taken many times to the Detroit Model Railroad Club, located across from the Grand Truck Railroad Depot on Detroit's riverfront, in order "to see the trains." When my first grandchild arrived in the 1980s, I began to set up an electric train set beneath the annual Christmas tree. Bob kept saying "I want to help next time." Unfortunately he never made it, something always intervening. More importantly, Bob, myself, and Jim Warfield (in the University of Illinois School of Architecture) each came to teach courses in what came to be called "vernacular architecture."

I cannot help but think that Bob's growing up in transient hotels impacted his sense of belonging. His was a childhood of living not in a house, but in a hotel. He lived not in a fancy suburb, but in rapidly changing inner-city locales. It must have marked his personality somewhat. And then there was the issue of his not having an advanced academic degree (as if any colleague cared). With academic professionals (and certainly with the many visitors he entertained on campus), he had a tendency to be always "on stage." He tried, and indeed succeeded, in rarely saying or doing a wrong thing it seemed. Perhaps, like Emily Post (who grew up in many of the hotels that her architect father had designed), Bob became something of a connoisseur of public social behavior. He always seemed to be "on guard." For many years I wondered about our never having become really close friends. There always seemed to be something of a distance between us. And then I figured out the above. I also came to realize that his manner of being in the world was very much the secret to his many successes: for example, as a promoter of academic programs and as a leader in professional organizations. And as a dedicated editor. He was the "go-to-person" whenever something innovative, important, or otherwise challenging needed doing, and doing right.

Bob never published much, yet his essays are powerfully insightful. What he wrote always seemed vital and absolutely timely. He always gave his writing very careful thought. He had a way of taking even the most mundane of topics to a high level of interpretation, for example: "pigs upwind" as a means of understanding the nature of Midwestern farmstead location, the landscapes of the rural Midwest being a favorite intellectual focus. Some might say he should have published more. But that could be said of all of us whose business it has been to interpret the American scene (if not scenes worldwide) in terms of evolving landscape and place.

Riley at home with weaving, ca. 2012.
(Photograph courtesy Illini Studio)

Carol Emmerling

I had the good fortune to be both a student and a colleague of Robert Riley. When I entered the MLA program at Illinois, and asked for advice, students told me I had to take at least one class with Bob . . . and I did. When I learned of Bob's death, I made a list of things I remembered about him. Every time I wrote something, I thought of something else . . . and that something else was usually in contrast to the first. Bob was both an architect and a Fellow of the American Society of Landscape Architects. His elevation to Fellow was in recognition of his service to the profession, specifically as a department head who shaped one of the premier landscape architecture graduate programs in the United States.

Bob was urbane and well-traveled, but he was a Midwesterner. In his Cultural Landscape course, we read *New Yorker* articles by Joan Didion *and Reading the Landscape of America* by May Watts. I still remember the first time he referred to Mies. Coming from Chicago I was suitably impressed. I soon learned that, in contrast to the architecture of the Bauhaus, Bob documented the vernacular landscape through his evocative photos.

His home was like a gallery for Southwestern art, and he generously shared it with graduate students and faculty. The food was excellent and the conversation stimulating, but the wine was always outstanding. Bob was a superb teacher and an even better colleague. John Steinbeck wrote, "It is so much darker when a light goes out than it would have been if it never shown." I am grateful to Bob for being that light.

Sue Weidemann: "Hogs to the East": Things That a Prairie-born Woman Learned from a Big City Man about Life on the Farm

"Square to the road" . . . another observation from Robert Bartlett Riley as we drove through the central Illinois landscape looking at the rural houses. I'd never heard such words, nor had I ever thought about why they were important, even though I had lived with what they described my whole childhood and life before the 'big U,' otherwise known as the home of the Fighting Illini (no longer appropriate), but officially the University of Illinois.

He, the architect, offered me (the psychologist) my first teaching job . . . in the Landscape Architecture Department there at the U of I. Contradictions? Complexities? Yep. But I didn't even think twice about it. Of course!

Bob will always be there in my memories, and in the memories and experiences of many others. He opened the eyes and minds of so many people to so many things. Curious, exploratory, fascinated with rural features, landscapes, and buildings . . . really "smart" as we modest prairie folk sometimes say to describe someone we can't really describe. But always exceedingly interesting. Well, except when he was cranky; but even that was interesting. Ha! A man of many complexities.

But we all learned from him. And I learned, although I didn't practice it nearly enough, to really "work hard" to produce something worthwhile. For those of you who might have asked him for a reference? If he agreed, I'll bet you never realized how long he spent, how carefully and thoughtfully he worded the document, so that your very best characteristics would be revealed to the reader.

And then there were the drives in the rural landscape just to look at things and talk about "what" and "why" . . . with Terry Harkness, John Jakle, and Bob. That was a combo! I could just sit in the back seat and get the best free education that anyone could ever imagine.

And a drive in the country to show a new job candidate around? What? Out in the country? What were we thinking? Well, it was a bit of an unrecognized test. If we stopped on an empty gravel road, surrounded by twelve-foot-high cornfields and endless rows of soybeans under the biggest blue sky you could image, and got out of the car to "look around"—then came the test. If the visitor started to get out, and then kind of hunched over and looked up and around with a worried look, we were pretty sure they wouldn't make it at Illinois. (That's called an "open-field test"; look it up.)

But for those of us who were there at Illinois, with Bob . . . all of us learned . . . to think about things in new ways, with curiosity. What joy!

Kenneth Helphand

We almost only met at conferences. I looked forward to it, for there was no one who was more fun or stimulating to talk with. At CELA, Dumbarton Oaks, or EDRA we invariably found ourselves sitting in or being participants in the same sessions. I was a follower of Bob, not in the sense of being an acolyte, but more literally. He was chair of the Senior Fellows in Garden and Landscape Studies at Dumbarton Oaks, I occupied the same position some years later. He edited *Landscape Journal;* I followed him as the next editor. But most important was that we both followed the landscape path laid before us by J. B. Jackson. Working and teaching in New Mexico Bob met Jackson and worked with him as associate editor on *Landscape*, the incredible magazine that Jackson founded. I was fortunate enough to be one of Jackson's teaching assistants when I studied landscape architecture at Harvard. Inspired by Jackson we both wrote and taught about ideas and places in the ordinary American landscape, and we both studied, taught, and wrote about gardens. Which brings me to my favorite Robert Riley story. It was 1985 and we were in a conference session at CELA at Illinois talking about gardens. I gave my presentation and sat down next to Bob. He then looked directly at me and grumbled "Son-of-a bitch, Son-of-a bitch." I felt horrible. Here was a man who I both liked and admired cursing me out. It took a few minutes, but then Sue Weidemann, his colleague in the Landscape Architecture Department at Illinois, could see the expression on my face, and she turned to me and said, "Kenny, it's his highest compliment." The paper I gave was entitled "The Garden." It had ideas about the meaning of gardens that festered in my mind for many years, ultimately culminating in my book *Defiant Gardens: Making Gardens in Wartime.*

Suzanne Turner: Bob Riley and Me

A neophyte, I was beginning my life as a landscape architect and teacher, and was seeking kindred spirits. I was thrown into CELA leadership by hosting an annual meeting of the teachers' group. There I first encountered Robert Riley: provocative, challenging, rigorous in his critique, but also funny, delightful, and good company; he made a routine board meeting an adventure. Perhaps more than anyone before or since, Riley believed that serving the profession mattered, and his impact in leading ASLA, CELA, *Landscape Journal*, and *Landscape* demonstrated that.

I would eventually succeed Riley as president of CELA and learn more about his gifts, among them his innate kindness and concern in helping young teachers and scholars like myself learn the ropes. It was this caring nature and sincerity that drew us all to him. Unlike other larger-than-life figures of the profession in the 1970s–1980s, Riley was approachable, genuinely interested in his colleagues.

What I most respected was Riley's mind—his ability to take an idea and develop it in an unexpected direction. The depth of his interests and observations was the platform that allowed him to pivot so freely in shaping ideas. His interest in everyday people and places was our common ground.

When Riley was asked to review my dossier for tenure, he not only supported my case, but sealed the deal by saying something like, "if this kind of scholarship, teaching and service does not meet your criteria for tenure, there are many other programs where Turner would be most welcome to join with tenure." I have been forever grateful and touched by his support.

Misa Inoue

Bob was my professor at the University of Illinois at Urbana-Champaign.

As I was preparing this writing, I wondered if there was one memory that I could single out, but I soon realized that I could not do justice if I chose just one. Everything I remember about Bob is too precious to be numbered.

I feel that I am writing to represent all my classmates who knew Bob. He was an incredible intellectual who could discuss any topic, and although he was our professor, at times it felt as if our roles had reversed. Whenever a student said something that piqued his interest, his face lit up and he suddenly took on the air of a student himself. What was so great about him was that he always stayed open-minded at whatever we said. He had such a big heart.

As a true educator full of curiosity and strong sense of fairness, he was also witty and thoughtful. By the time I returned to the school as a visiting practitioner several years ago, he had retired from teaching, but whenever I asked, he made the time to visit my class to offer critiques. Although throughout his life he had been a towering figure at and beyond the university, he had this amazing ability to connect with every student at their own different stage of learning. For me, to witness these exchanges between him and my students was a tremendous honor.

He had great style, and never said or did anything half-heartedly. And, when there was something that he did not like or agree with, he had little hesitation to let it be known. He was always very honest.

I am forever grateful that I had met and learned with him. There is no other like Bob.

Gary Hilderbrand

As a young teacher, I had the good fortune of spending time with Bob during a semester he spent at Harvard teaching a seminar on landscape photography in the late 1990s. I loved that he insisted on the importance of John Dewey's pragmatism in breaking down Robert Frank's *The Americans* for landscape architecture. I secretly wanted to break down the contrasting image of his dour brown sport jacket and flashy turquoise belt buckle. I think he probably would have been pleased at my complete inability to solve this contradiction. Big presence, big mind, big man. Big impact.

Joachim Wolschke-Bulmahn

I could experience Bob as editor of *Landscape Journal* and as a member of the board of Senior Fellows at Dumbarton Oaks when I was director of Garden and Landscape Studies there. I appreciated Bob very much as an editor because he was open for 'uncomfortable' topics, among them German landscape architecture and National Socialism. And as member of the board of Senior Fellows he was refreshingly different than the other board members—a creative lateral thinker.

When I took over the directorship at Dumbarton Oaks in 1991 that was the starting point for a closer connection to Bob. Based on the holdings of the garden library, I wrote a paper, "From the War Garden to the Victory Garden: Political Aspects of Garden Culture in the United States during World War I," which Bob published in *Landscape Journal* in 1992. Shortly later a paper, jointly written with Gert Gröning, "Some Notes on the Mania for Native Plants in Germany," was also accepted for publication in *Landscape Journal*. This article created a heated debate. That Bob accepted it for publication demonstrates his willingness to publish provocative articles and open *Landscape Journal* for political discussion. After our article's publication Bob published letters—sometimes lengthy—in response to it. Two titles suggests the discussion's character: "Natives and Nazis: An Imaginary Conspiracy in Ecological Design. Commentary on G. Groening and J. Wolschke-Bulmahn's 'Some Notes on the Mania for Native Plants in Germany'"); and Gert's and my response, "If the Shoe Fits, Wear It!" This indicates one of Bob's particular qualities as an outstanding editor—always willing to promote discussion, to open *Landscape Journal* for sometimes heated debates.

It was a great pleasure for me to suggest Bob to Angeliki Laiou, then director of Dumbarton Oaks, for the Senior Fellows Committee for Garden and Landscape Studies there; Bob accepted Angeliki's invitation and served on that committee from 1992 to 1999, the last three years as its chair. An important product of his work as enior fellow is the conference volume *Theme Park Landscapes: Antecedents and Variations,* co-edited with Terence Young and published as volume 20 in the series *Dumbarton Oaks Colloquia on the History of Landscape Architecture.* Work on the symposium began shortly before my return to Germany in 1996. After I left, Terence Young took over as acting director for a year and Bob was the senior, experienced co-organizer of the symposium.

David L. Hays

I first met Bob at Dumbarton Oaks in 1997, when he was chair of the Senior Fellows Committee in Landscape Architecture and I was a new junior fellow aiming to finish my dissertation. Four years later, I joined the faculty of Landscape Architecture at Illinois. Laura Lawson and Dede Ruggles were also new, and, as we walked together into our first faculty meeting, Bob turned and called out, "Thank God you're here!"

During the years that followed, I got to witness firsthand Bob's remarkable way of being with students and to learn from alumni about his transformative influence on their lives. He was a giant, a hero—and yes, his voice was amazing, both spoken and written. The latter is preserved in our archives, in documents from Bob's time as department head. A reply to a dean declares, "I sure as hell have no intelligent ideas for solution," though what follows is a page of careful speculations. A letter written in frustration begins, "For shame!" but, after measured explication of the matter at hand, it closes, "Peace."

At a department luncheon in May 2006, Bob and I sat next to each other. He was in very low spirits. Reading news to distract himself, he had learned about a website where you could answer questions about your habits and it would predict the date of your death. He decided to do it. Do you exercise? How often do you socialize? How much do you drink? Do you feel depressed? He answered dozens of questions, held his breath, and hit submit. The answer came back: February 10, 1987.

I burst out laughing.

"But what am I supposed to do with *that*?" he demanded.

"Bob," I cheered, "*don't change a thing!*"—and suddenly, we raised and touched our glasses in a merry toast.

Doug Johnston

I first met Bob when I was a candidate for a faculty position at Illinois. For a young applicant, with no significant experience, he was an imposing, if not terrifying presence. Certainly not the kind of department chair I expected. But his intellect and persona indicated that working for him would be challenging and interesting.

To me, he was the epitome of "cool." Casual jeans, crisp white dress shirt, and of course, the enormous belt buckle. Porsche, custom-designed home. Loud, opinionated, forthright.

As a senior member of the faculty, he was a gracious host, whether meeting with him at his corner table at Timpone's for a glass of wine, or at a party at his house for many glasses of wine. Rooms and hallways lined with books, and the most unusual and interesting Southwest art always made for fascinating meanders and conversations.

On more than one occasion he told me I needed to be more selfish, to say no occasionally, but he was damn generous.

Frederick Steiner

The Council of Educators in Landscape Architecture (CELA) played a significant part of my scholarly life and Bob was my unofficial CELA mentor. Like Bob, I came to landscape architecture education sideways: he, as an architect, through J. B. Jackson, and me, as a planner, through Ian McHarg. Through Bob's guidance and example, I became a better landscape architecture educator. Smart, funny, encouraging, and engaging, Bob elevated CELA as an organization: He was unwavering in his commitment to raising the level of scholarship and to increasing diversity in the discipline. He was an especially strong and effective advocate for women.

My first impression of Bob was that he was a grouchy cowboy who looked a bit like Steve McQueen with a goatee. Beyond the rugged, serious demeanor was an urbane, erudite gentleman and scholar. Like his mentor, Bob was especially attached to ordinary American landscapes. Like Jackson, Bob was informed by a broad knowledge of the history of the built environment and international travel.

Chip Sullivan

Unlike "Brinck" (I learned from Bob that's what Jackson's friends called him), Bob was more of an organization man. He assumed leadership roles in his department at the University of Illinois, in CELA, and in the American Society of Landscape Architects. He organized meetings and conferences. I remember a CELA conference in Champaign-Urbana that coincided with the first Farm Aid concert. I flew to Chicago in a small plane in a thunderstorm with the Beach Boys with the proceedings Bob edited on my lap.

Bob was always deeply present at events. He sat near the front even at the most junior scholar's presentation and found a good question to ask. He ran meetings that encouraged engagement with a refreshing lack of hierarchy.

Bob was an inspiration.

Reuben M. Rainey: A Dinner with Bob

If you ever needed someone to critique your latest idea for an article or point you toward some new territory for research, you could do no better than a one-on-one conversation with Bob Riley. I vividly recall my first encounter with Bob in the early 1980s over one of those rubber chicken, suspect Chardonnay, and petrified chocolate cake dinners at a conference in Chicago. We quickly discovered we had similar backgrounds in philosophy, exposure to a great books curriculum, and the teaching of history, as well as admiration for J. B. Jackson, good wine, and German sports cars (we drove the same model). Then we ventured into deeper conversational waters. We focused on how best to teach history of landscape architecture and the relationship between environmental psychology and design. I emerged from that conversation with a research agenda that occupied me for many years and an acute awareness that I needed to dispense with my pretentious design babble. I am certain many others have had similar experiences, and I was fortunate to have occasions to talk with Bob over many years with similar results. The following quotations from Bob's astute, wise, blunt, and thought-provoking reflections are but a few that have catalyzed the thinking of many practitioners and academics for a generation. They are taken from his collection of essays, *The Camaro in the Pasture.*

> ". . . history must be relevant to design. It must make the student reflect upon decisions that she makes in the design process." (p. 34)

Riley at design reviews, University of Illinois, 2011. (Photograph by Misa Inoue)

"Our canon is biased and arbitrary. Whose is not? Our job is to encourage our students to ask questions of it." (p. 34)

"The essence of landscape is change, a fact we find hard to admit." (p. 35)

"If I were granted one wish it would be for a renewal of the eighteenth- and nineteenth-century debate on the nature of the landscape experiences, updated with the tools and the vocabulary of the social and behavioral sciences of the twenty-first century." (p. 145)

Randolph Hester: Bob Riley

For his mastery of stock car racing and daring moves Dale Earnhardt was monikered the "Intimidator." If you got in his way on the track, Earnhardt ran you over, raising his sport to a more exciting level. Similarly, Bob Riley was a master of the field and intellectually intimidating. With daring moves of his mind, he often ran over ill-formed ideas, advancing our profession. Acutely philosophical and complexly read, Bob thought deeply and irreverently. He held a high-minded view of the vernacular, challenged sacred canons, made the familiar strange and the strange familiar. He championed irreconcilable oppositions. He turned fashionable gardens into designer jeans. He turned phrases like Earnhardt took corners.

He was a treasured friend, but Bob was intimidating. He was no brainy bully. He never stooped to imperious tactics. But his thoughts shook me like a championship engine doing victory burnouts. Earnhardt once said after wrecking a competitor, "I just wanted to rattle his cage." Bob would enjoy the comparison. He too rattled our cages, caressing his goatee and smiling with appreciation.

He compelled, yeah even frightened, me to speculate, study, reflect, and design differently. Bob had a witty, spell-binding, and thought-provoking way of poking me in the eye. While informing me of some important idea his dazzling humor always made me laugh, often at my own expense. He had to suffer a lot of design fools, but he encouraged me to do better, to question comprehensibly, to reason critically, to search for inspiration in unexplored places. He taught anyone who listened to grasp landscape design with rational breadth and expressive imagination. The theoretical foundations of our profession progressed immeasurably from his efforts. More than anyone else of his era Bob Riley raised the intellectual bar of landscape architecture.

Afterword

When I learned Bob Riley had died, I flailed about for some time, searching for what I might do—as if action could fill the gaping hole left behind. Eventually I decided that the one thing that made sense, the one thing I knew how to do, was to create a book in his honor, a book with many voices, because Bob Riley touched many different lives, often in profound ways. Perhaps this choice should not be surprising. While my relationship with Bob had many facets and extended over thirty years, it was forged in those years when he shared his magisterial Mumford Hall office with me and we worked together on *Landscape Journal*—my assistant editor to his editor. Whatever predilections I may have brought to that job, it was then that I learned about being an editor and assembler, about bringing together varied players, and about some of the potential results.

Backward looks were unavoidable, and it occurred to me I might be trying to recreate the sort of *Landscape Journal* on which I worked with Bob, but it was important to me that this book not be an exercise in nostalgia, neither my own nor that of others. I wanted—and still want—it to be inspiring, especially for thoughtful, younger people concerned with landscapes and their design today.

I invited contributions to this book—both reminiscences and essays—based on what I knew of Bob's interactions and influences. Surely there are gaps due to my partial knowledge, personal comfort, and others' responsiveness, especially in 2020, the pandemic year in which these essays were first written and much of the book assembled. In some cases, I felt very confident to whom I reached out; sometimes I followed a hunch or an aside Bob had made. A few times, for various reasons, people did not respond; sometimes I was moved to discover new aspects of Riley's influence. I hope anyone who feels left out will understand it is due to my limited perspective.

It was also important to me that younger contributors—writers and scholars unlikely to have known Bob—be amongst the essayists. Together with the project assistant, Kathryn McCudden, I made a concerted effort to find appropriate writers who, we surmised, were under forty—or close to it. Bob, I thought, would be depressed if all the contributors to such a work were considerably older than that. I am aware that female voices predominate in the essays as a whole. This was not intentional but when I noticed it I did not fight it.

I think the Riley essays herein are among his best. Some I planned to include from the book's conception; some found their way in because the essayists responded to a different Riley work as they crafted their own. However, it should not be thought that Riley's concerns and insights were limited to these topics or that these were the only times Riley wrote about these subjects. As one can read in his *The Camaro in the Pasture*, place attachment, landscapes and memory, and landscapes in literature—including landscapes of science fiction—were just a few of his other subjects. (Indeed, his description of a future landscape in "The Goose and the Dish" may owe something to the science fiction book covers in his slide collection.)

Reflecting on his career, Bob once wrote:

> I think mostly what I have done is raised issues, doubts and speculations about topics that are important to me and have often turned out to be important for other people. . . . As a writer and a thinker, I'd like to be remembered as somebody who injected plain language and common sense into Landscape Architectural debate without lowering intellectual standards. . . . I don't know that what I've helped students learn has made them design better. I'm not even sure if they've designed differently because of what we've explored, but I do know, because of what I've been told, that I've helped some of them *look at the world differently.* Not to be morbid, but I suppose that wouldn't be a bad epitaph, would it?[1]

At many conferences and symposia (and similarly for publications), Bob would be the one called upon to make summary comments and articulate smart, pithy takeaways. He was given such tasks because he was good at it; he listened well and could make connections, recognize patterns, sense trends, and identify gaps, and he presented these in ways that, as Beth Meyer has put it, elicited "gasps and giggles." He could do this—seemingly—on the fly (he could think on his feet), as well as later, more reflectively and extensively, in his writing.[2]

Another of Bob's great strengths was his aptitude and willingness to seek out, probe, absorb, weigh, evaluate, and question what was going on around him. Terry Harkness, who of all those included here knew Bob the longest, may have put it best.

> Bob wanted to understand landscape and design and so he would seek out and really listen to people involved in architecture, design, and planning. And then he would think about it critically . . . testing, weighing, questioning. . . . Bob was so unusual in his curiosity and his large view of the world. . . . He had a much richer and interesting world that was looking outward, not just about what landscape architects thought about or what architects thought about. I just thought that was a kind of magical quality.

Bob's was far from a tunnel vision. I often marveled at the genuine curiosity and openness he combined with an ability and willingness to cut through the crap, crap in the way of intellectual pretensions and limiting frames of reference, but also institutional, social, and personal quagmires, hence the many people who came to him for advice. (He once remarked on the treacherous path between wise sage and old fart.) Bob's sense of the larger world was not limited to the current popular mantra, however true, that all disciplines must work together to solve the world's wicked problems. Nor was it based on superficial forays into others' specialties. He did not formulate popular theories; rather he took pains to distinguish theories, models, and frameworks, and was perhaps particularly adept in constructing the last. He did not write books on a single topic, he wrote essays, some very long and, on occasion, others pointedly short.

The reminiscences and assessments included here convey a lot about Bob's personality and influence. In one way or another all the essays here demonstrate the continuing relevance and importance of Riley's larger view and way of looking at the world. He was a designer but perhaps more singularly an erudite humanist and lover of landscapes, one fascinated by and provoking others to consider and reconsider landscapes, design, and how we and others see, experience, and think about them.

I miss Bob Riley a lot. Working on this volume has at times made that missing and regret for conversations deferred and words not spoken especially acute. Ironically, he is the one with whom I would have most liked to talk about the pleasures and vicissitudes in its process. I ponder now what Bob would make if it. I hope he would like it. I hope it would stir him to new reflections, ideas, and insights. He so enjoyed that, and in his enjoyment—however curmudgeonly—stirred others to do the same.

NOTES

1. Robert B. Riley, "Educational Accomplishments," unpub., n.d.
2. At this book's end, faced with an analogous task, regretfully, I claim no such similar skills.

Acknowledgments

I am deeply grateful to all the contributors to this volume—the authors of the essays as well as all those who shared their own reminiscences of and perspectives on Robert B. Riley and his work. Each rendered encouragement for the project. Together they provide a much fuller and more nuanced portrait of Bob than would be otherwise possible.

I was extremely lucky to have Kathryn McCudden as assistant for this project, first when she was a Master of Landscape Architecture student at the University of Manitoba, and then, continuing on after her graduation and employment in Regina, Saskatchewan. An Olmsted Scholar, Kathryn became involved soon after the project's start. She quickly familiarized herself with Riley's writings and, accordingly and sensitively, helped identify younger potential contributors, provided thoughtful feedback on essay drafts, and saw to much of the meticulous attention necessary in this endeavor, including the assembling of its index.

While I originally imagined this as a special journal issue, Elen Deming suggested its viability as a book. She introduced me to Lake Douglas, who sympathetically helped shape it to become part of his Reading the American Landscape Series at Louisiana State University Press. Jenny Keegan, my main contact at the Press itself, suffered my impatience and penchant for details with upbeat grace. I much appreciated Catherine Kadair's informed graciousness, copyeditor Jo Ann Kiser's keen eye and sensitive touch, and designer Michelle A. Neustrom's receptivity and willingness to listen and communicate with me directly. I thank Rebecca H. Riley, Bob's daughter, for various information and support needed along the way, as well as, in her role as executor of Robert B. Riley's estate, her permission to reprint Bob's essays and to use images from the family collection and the University of Illinois Ricker Art and Architecture Bob Riley Landscape Architecture Collection.

Some of the essays included here also appeared, sometimes slightly revised, in Riley's *The Camaro in the Pasture: Speculations on the Cultural Landscape of America* (University of Virginia Press, 2015). I have chosen, though, to reprint them as originally published. I hope they will thus be better considered in relation to the times when they were written as well as provoke reflection on how little major issues and concerns have changed. I gratefully acknowledge the following for permission to reprint Robert B. Riley's previously published essays from those original publications: "Editorial Commentary: Some Thoughts on Scholarship and Publication," *Landscape Journal* 9, no. 1 (1990): 47–50, © 1990 by the Board of Regents of the University of Wisconsin System, reprinted courtesy of the University of Wisconsin Press; "What History Should We Teach and Why?" *Landscape Journal* 14, no. 2 (1995): 220–25, © 1995 by the Board of Regents of the University of Wisconsin System, reprinted courtesy of the University of Wisconsin Press; "Square to the Road, Hogs to the East," *Places Journal* 2, no. 4 (1985): 72–79; "From Sacred Grove to Disney World: The Search for Garden Meaning," *Landscape Journal* 7, no. 2 (1988): 136–47, © 1988 by the Board of Regents of the University of Wisconsin System, reprinted courtesy of the University of Wisconsin Press; "Riley's Auto Territoriality," *Landscape*, Spring 1968, pp. 1–2, MSS 633 BC, J. B. Jackson Papers, Center for Southwest Research, University Libraries, University of New Mexico; "The Goose and the Dish" in *The Camaro in the Pasture: Speculations on the Cultural Landscape of America*, 154–57, © 2015 by the Rector and Visitors of the University of Virginia, reprinted by permission of the University of Virginia Press.

Thanks to Kyle Rimkus, Preservation Librarian at the University of Illinois at Urbana-Champaign, for help in navigating the Bob Riley Landscape Architecture Collection and to Patrick Kisel at Luftbild-Bertram for finding one of my favorite of Riley's favorite photographs (and to Dietmar Straub for ferreting out the German citations to get me to him). Thanks also to Terry Harkness, Misa Inoue, Bernard Lassus, Paul Lettieri, Stephen Martiniere, Michael Van Valkenburgh, Kenneth Helphand, and Susan Wydick for use of their images.

Parts of my own brief biography of Bob in this book's introduction previously appeared in "A Tribute to Robert B. Riley," *Landscape Journal* vol. 38 (2020): 1–2, © 2020 by the Board of Regents of the University of Wisconsin System, reprinted courtesy of the University of Wisconsin Press.

Finally, heartfelt thanks to Phil Boyland for keeping track of me as I kept track of this project, and thanks to Bob Riley, through whose capacious and personal love of landscapes, thought, and work I found the realm of my own.

Contributors

BRENDA J. BROWN is associate professor of landscape architecture at the University of Manitoba. She was assistant editor of *Landscape Journal* from 1990–95, when Robert B. Riley was editor. She chaired the award-winning Committee for Eco-Revelatory Design that produced the Eco-Revelatory Design: Nature Constructed/Nature Revealed exhibition and catalog of the same name, a special issue of *Landscape Journal* for which she was editor. Her publications include articles in *Landscape Journal, Landscape Architecture, Landscapes/Paysages, Journal of Landscape Architecture, Ecological Restoration and Center,* and chapters in several edited volumes. Her lavishly illustrated bilingual exhibition catalog *Tzintzuntzan, el lugar de los colibríes—otra vez / Tzintzuntzan, Place of the Hummingbirds—again* (2015, for which she was editor, included contributions from eleven writers of diverse backgrounds. Her landscape and multimedia installations have been heard and seen in Canada, the United States, and Mexico.

JAMES CURTIS studied with the master gardener Kinsaku Nakane in Japan on a Ryerson Traveling Fellowship after graduating from the University of Illinois. He has worked for POD, Inc., and EDAW—EDAW | AECOM with entr'actes at Michael Van Valkenburgh Associates, Peter Walker William Johnson Partners, and Ken Smith Workshop West. He now manages the Landscape Studio in the Irvine office of WATG and teaches and serves on the guidance committee of the UCLA Extension Landscape Architecture Program.

M. ELEN DEMING has been a practitioner and educator in and around landscape architecture for nearly forty years. Her vantage on research trends in the field is informed by her professional and academic training, editorship of *Landscape Journal* (2002–2009), research leadership for the Landscape Architecture Foundation (2016–2019), and, now at North Carolina State University, administration of a unique doctoral program in transdisciplinary design research (since 2017). She is coauthor of *Landscape Architecture Research: Inquiry, Strategy, Design* (2011), and editor of *Values in Landscape Architecture: Finding Center* (2015) and *Landscape Observatory: The Work of Terence Harkness* (2017).

CAROL EMMERLING is a landscape architect. She taught in the Department of Landscape Architecture at the University of Illinois for twenty-nine years.

ROSA E. FICEK is a cultural anthropologist at the Institute of Interdisciplinary Research at the University of Puerto Rico at Cayey. Her work addresses technology, empire, and the environment in Latin America and the Caribbean. She is a recipient of a Smithsonian Institution Fellowship and a past fellow in Dumbarton Oaks' Garden and Landscape Studies. Her articles have appeared in *Current Anthropology*, *Transforming Anthropology*, and the *Journal of Transport History*.

SHARON HARKNESS has a bachelor of science degree and ten years' experience in nursing. Upon returning to Champaign-Urbana with husband Terry in 1981, she began working in real estate, and she continues to do so.

TERRY HARKNESS is professor emeritus of landscape architecture at the University of Illinois at Urbana-Champaign. A designer's designer, he has won major awards for his landscape architecture as well as for his teaching. He was a junior member of the University of Illinois's landscape architecture department when Robert Riley was hired, and returned after eight years in practice to be Riley's colleague for the next twenty-six years.

DAVID L. HAYS is the Brenton H. and Jean B. Wadsworth Head of the Department of Landscape Architecture at the University of Illinois at Urbana-Champaign.

KENNETH HELPHAND is the Philip H. Knight Professor Emeritus of Landscape Architecture at the University of Oregon. He is the author of *Colorado: Visions of an American Land-*

scape (1991), *Yard Street Park: The Design of Suburban Open Space* (with Cynthia Girling, 1994), *Dreaming Gardens: Landscape Architecture & the Making of Modern Israel* (2002), *Defiant Gardens: Making Gardens in Wartime* (2006), and *Lawrence Halprin* (2017).

RANDOLPH HESTER is professor emeritus at the University of California, Berkeley; director of the Center for Ecological Democracy in Durham, North Carolina; and founder of the Shorty Lawson Museum of the Black Tenant Farmer in Hurdle Mills, North Carolina.

GARY HILDERBRAND is principal of Reed Hilderbrand and the Peter Louis Hornbeck Professor of Practice in Landscape Architecture at the Harvard Graduate School of Design. He's a big fan of Bob Riley.

LEWIS D. HOPKINS is professor emeritus and former head of the Department of Urban and Regional Planning, University of Illinois at Urbana-Champaign. Hopkins was Robert Riley's first hire as head of the Department of Landscape Architecture at the University of Illinois in 1971. His books include *Urban Development: The Logic of Making Plans* (2001) and the coedited volume, with Marisa A. Zapata, *Engaging the Future: Forecasts, Scenarios, Plans and Projects* (2007).

MISA INOUE obtained her MLA at the University of Illinois at Urbana-Champaign. She is a landscape architect currently based in Chicago, Illinois.

JOHN JAKLE is emeritus professor of geography at the University of Illinois at Urbana-Champaign. He has authored and coauthored books focused on a range of diverse but nonetheless interrelated topics including the automobile's impact on the American built environment, American vernacular architecture (common housing and especially roadside commercial architecture), nighttime lighting and the geography of night, the depiction of landscapes and places in postcard art, and the American small town as landscape and place.

DOUG JOHNSTON served on the faculty in the Department of Landscape Architecture at the University of Illinois at Urbana-Champaign from 1986 to 2007. From 2007 to 2013 he was professor and chair of the Department of Landscape Architecture and the Department of Community and Regional Planning and interim associate dean for research in the College of Design at Iowa State University. He is now professor and chair of the Department of Landscape Architecture and interim director of the Open Academy at the State University of New York (SUNY) College of Environmental Science and Forestry (ESF).

RACHEL LEIBOWITZ is assistant professor in the Department of Landscape Architecture and codirector of the Center for Cultural Landscape Preservation at the State University of New York, Syracuse. Her articles have appeared in *Buildings and Landscapes: The Journal of the Vernacular Architecture Forum*, *Arboriculture and Urban Forestry*, and *Planning Perspectives*. Her current book project is entitled "Constructing the Navajo Capital: Landscape, Power, and Representation at Window Rock."

REUBEN M. RAINEY is William Stone Weedon Professor Emeritus in the School of Architecture at the University of Virginia.

ROBERT B. RILEY was professor emeritus of landscape architecture and architecture at the University of Illinois at Urbana-Champaign. He was editor of *Landscape Journal* (1987–1994), and his publications appeared in numerous journals including *Landscape*, *Landscape Journal*, *Places*, *Harvard Design Magazine*, and *Landscape Architecture*, as well as in edited volumes such as *The Landscape Approach of Bernard Lassus* (2000), *Understanding the Ordinary Landscape* (1997), and *Place Attachment* (1992). His book *The Camaro in the Pasture: Speculations on the Cultural Landscape of America* was published in 2015.

ACHVA BENZINBERG STEIN is professor emerita and founding director of the graduate program in landscape architecture at the Spitzer School of Architecture, City College of New York, where she was a member of the faculty in both landscape architecture and urban design. She designed the Moroccan Courtyard at the New York Metropolitan Museum and has received many awards for her teaching and professional work. Among her publications are the book *Morocco: Courtyards and Gardens* (2007) and essays in the edited volumes *Historic Cities and Sacred Sites/Cultural Roots for Urban Futures* (2001) and *The Next Jerusalem: Sharing the Divided City* (2002). Stein has taught and worked in the United States, Israel, Europe, India, and China and is the recipient of two Fulbright fellowships—one to India

(1979) and one to Germany (2001). She continues to work on special projects in the US and abroad.

FREDERICK STEINER is the dean and Paley Professor of the Stuart Weitzman School of Design at the University of Pennsylvania. A fellow of the American Academy in Rome, the American Society of Landscape Architects, and the Council of Educators in Landscape Architecture, he has written or edited twenty books about planning and design.

CHIP SULLIVAN is an artist and professor of landscape architecture at the College of Environmental Design, UC Berkeley. Chip received the 2016 Jot D. Carpenter Teaching Medal from the American Society of Landscape Architects, which recognizes excellence in landscape architecture education. His latest book, *Cartooning the Landscape,* concerns the metaphysics of drawing and learning how to "see."

LINNAEA TILLETT is principal and founder of Tillett Lighting Design Associates. An environmental psychologist as well as a lighting designer, Tillett has worked with many of the leading designers in North America, including Olin Partnerships, Piet Oudolf, Maya Lin, Michael Van Valkenburgh Associates, Civitas, and Gustafson Porter Landscape Architects. She is known for her social and ecological awareness as well as for the subtlety of her lighting designs. She has lectured at universities and conferences all over the Western world.

SUZANNE TURNER taught in the Louisiana State University School of Landscape Architecture for twenty-seven years, holding various positions including graduate program coordinator, interim school director, and associate dean of the College of Design. Turner's career has focused on the study of cultural and historic landscapes; she is particularly interested in their interpretation and preservation. She has authored several books and numerous articles and has lectured widely.

MICHAEL VAN VALKENBURGH is president and CEO of Michael Van Valkenburgh Associates (MVV Associates), one of the preeminent landscape architecture firms in North America today. He is also former Charles Eliot Professor in Practice of Landscape Architecture at the Graduate School of Design, Harvard University.

Index

VERA VICENZOTTI is senior lecturer in landscape architecture at the Department of Urban and Rural Development of the Swedish University of Agricultural Sciences (SLU). Her research interests are landscape architecture theory, methodology, and history. An editor for the *Journal of Landscape Architecture,* she has contributed articles to that publication as well as to many others, including *Nature and Culture, Landscape Journal, Environmental Values,* and *Landscape Research.*

SUE WEIDEMANN is an environmental psychologist who has studied the relationships between people and the places and spaces they use for more than forty years. She taught for twenty-five years in the Department of Landscape Architecture at the University of Illinois. In 1994, she joined BOSTI Associates as director of research, and since 2011 she has been part of the IDeA Center at the University of Buffalo, where she also taught in the Department of Architecture. Since her retirement in 2021 she has been active in landscape restoration.

JAMES WESCOAT is Aga Khan Professor Emeritus at the Massachusetts Institute of Technology and former head of the Department of Landscape Architecture at the University of Illinois at Urbana-Champaign.

JOACHIM WOLSCHKE-BULMAHN was a fellow in landscape architecture at Dumbarton Oaks, a research institute at Harvard University, in 1989–1990. From 1991 to 1996 he was director of Studies in Landscape Architecture at Dumbarton Oaks. From 1996 to 2021, he was professor of the history of open space planning, Faculty of Architecture and Landscape Sciences, at Leibniz University Hannover.